T0260417

Spatial Structures

Originally published in 1973, this book synthesizes the mass of material into an introduction to the study of spatial systems. Geographic literature of the time stressed the influence of the distance between places on both location decision-making and movement patterns, arguing that the spatial system is an ordered set of interacting locations. This system is created by human decisions, influenced by the distance factor, and the system's morphology constrains further activities, including those which would alter it. *Spatial Structures* outlines the development of such systems, their present organization, and the ways in which they are changing. These themes are dealt with in three main chapters which focus on different spatial scales – the individual city, the nation state and the international system, within a simple classification of spatially organized activities.

Spatial Structures
Introducing the Study of Spatial Systems in Human Geography

R.J.Johnston

Routledge
Taylor & Francis Group

First published in 1973 by Methuen & Co. Ltd

This edition first published in 2023 by Routledge
4 Park Square, Milton Park, Abingdon, Oxon, OX14 4RN
and by Routledge
605 Third Avenue, New York, NY 10158

Routledge is an imprint of the Taylor & Francis Group, an informa business

© 1973 R. J. Johnston

Publisher's Note
The publisher has gone to great lengths to ensure the quality of this reprint but points out that some imperfections in the original copies may be apparent.

ISBN 13: 978-1-032-49307-7 (hbk)
ISBN 13: 978-1-003-39314-6 (ebk)
ISBN 13: 978-1-032-49315-2 (pbk)
Book DOI 10.4324/9781003393146

Spatial Structures

Introducing the study of spatial
systems in human geography

R. J. JOHNSTON

METHUEN & CO LTD

First published 1973 by Methuen & Co Ltd
11 New Fetter Lane, London EC4
© 1973 R. J. Johnston

Printed in Great Britain by Butler & Tanner Ltd
Frome, Somerset

SBN 416 76650 1 hardback
SBN 416 76660 9 paperback

FOR RITA

Contents

The Field of Geography

Progress in modern geography has brought rapid changes in course work. At the same time the considerable increase in students at colleges and universities has brought a heavy and sometimes intolerable demand on library resources. The need for cheap text-books introducing techniques, concepts and principles in the many divisions of the subject is growing and is likely to continue to do so. Much post-school teaching is hierarchical, treating the subject at progressively more specialized levels. This series provides text-books to serve the hierarchy and to provide therefore for a variety of needs. In consequence some of the books may appear to overlap, treating in part of similar principles or problems, but at different levels of generalization. However, it is not our intention to produce a series of exclusive works, the collection of which will provide the reader with a 'complete geography', but rather to serve the needs of today's geography students who mostly require some common general basis together with a selection of specialized studies.

Between the 'old' and the 'new' geographies there is no clear division. There is instead a wide spectrum of ideas and opinions concerning the development of teaching in geography. We hope to show something of that spectrum in the series, but necessarily its existence must create differences of treatment as between authors. There is no general series view or theme. Each book is the product of its author's opinions and must stand on its own merits.

W. B. MORGAN
J. C. PUGH

University of London,
King's College
August 1971

Preface

The main aim of the 'quantitative and theoretical revolution' in human geography has been to develop general theory concerning the spatial patterns of human activities. Yet, paradoxically, much of the recent literature in this field presents the results of specialized research projects, with no overall indication of their importance or position in the quest for theory. There is a great need for such a body of material to be collated into coherent wholes. This is attempted in the present book, which introduces the study of spatial systems within a framework which should prove useful in introductory undergraduate courses at universities and polytechnics, and also to teachers of the more advanced forms in secondary schools.

Such a brief introduction as attempted in this book cannot cover every aspect of a very wide area of study. The main focus is on the countries of the 'English-speaking' world, especially those considered the more 'developed'. Even within such terms of reference, certain topics must be virtually ignored; others are neglected. As a partial counter to this, the book is fully referenced, and a substantial bibliography is provided to lead the reader to the more detailed studies on which this introduction is based. One of the problems of using this literature for many readers is the language which it uses, both the jargon and the mathematics/statistics. This language is used where necessary in the present book. To help in its interpretation, a glossary has been provided, referring to terms marked by asterisks in the text.

Spatial Structures has a multivariate pedigree. Its contents reflect a number of courses which I have taught in recent years, most of them at the University of Canterbury. The present format was produced for a series of seminars in the Department of Geography at Monash University during October 1971, and I am grateful to that university, and to Professor Mal Logan, for the invitation. My friend and former colleague, Dr Joe Powell, encouraged me in that venture, and also to write down my ideas, stimuli which are much appreciated. Initial preparation of the material was undertaken at the University of Canterbury, using its excellent library resources and working in the fine academic environment of its Department of Geography. Once more, I am greatly indebted to Professor Barry Johnston for the stimulus of that setting, and for his encouragement of my attempts to develop courses using these materials. Much of the formal writing was accomplished while I was a visiting faculty member of the University of Toronto: I am indebted to Professor Don Kerr for the opportunity to teach in the Department of Geography there, to Professor Alan Baker in whose undergraduate course chapters 2 and 4 were tried out, and to Miss Susan Eberslie who typed a coherent manuscript from my handwritten scrawl. Finally, the book has been completed during my period as an Academic Visitor at the London School of Economics and Political Science, for which privilege I am indebted to Professor Michael Wise.

In preparing the manuscript, I am indebted to Miss Janice Price for her interest in the project and assistance in the many processes needed to complete it. As editor of the series, Professor W. B. Morgan has performed a marvellous task of transformation in the hours he has spent helping to polish a stuttering script. Much of the credit for the book's readability must be his, whereas of course the responsibility for any infelicities or errors is mine alone.

Finally, I am pleased to record my great debt to my wife Rita. This book is dedicated to her, not merely in recognition of all the help she gave in its preparation, but as one small token of thanks for ten years of dedication on her part to my work, and to the many hours when it has meant our separation. Without the environment which she has provided, none of this would have been possible.

March 1973   R.J.J.

Acknowledgements

The following permissions to reproduce diagrams are acknowledged:

The Association of American Geographers for fig. 2.2, taken from Wilson (1967)
Prentice-Hall Inc. for fig. 2.4, taken from Vance (1970)
Economic Geography for fig. 2.5, taken from Pred (1971a)
Geografiska Annaler for fig. 2.6, taken from Rimmer (1967b)
The Association of American Geographers for fig. 3.2, taken from Griffin and Preston (1966)
The Geographical Association of New South Wales and M. I. Logan for fig 3.3, taken from Logan (1964)
The Town Planning Review for fig. 3.5, taken from Lloyd (1965)
R. A. Murdie for fig. 3.6, taken from Murdie (1969)
Roy Chung for fig. 4.2, taken from Chung (1971)
The Geographical Review for fig. 4.3, taken from Zelinsky (1971).

Introduction

Human behaviour is affected and constrained in a great variety of ways. These can be grouped into four categories, or sets of environmental influences, those of: (1) the physical environment, the land, water and air man occupies, plus the fauna and flora he shares these with; (2) the social environment, the cultural, organizational and institutional structures erected by man as bases of his life; (3) the built environment, created by man; and (4) the spatial environment, the set of relative locations within which man exists. Each item of individual and group action is influenced by one or more of these groups of factors, acting separately or in conjunction. Understanding society involves unravelling these complexes of cause and effect, an endeavour often further complicated because behaviour takes place not within the 'real' environments but according to man's perceptions of those, from which they may differ considerably.

Most areas of scholarship are clearly identified with one of the four environments outlined here. Geography, on the other hand, has commonly studied the interactions among the environments. The classic regional synthesis, for example, embraced all four in its mapping of landscapes, emphasizing the first two and rarely attempting to identify causal relationships. The modern discipline of human geography concentrates much of its effort on two main sets of man–environment associations, those between the social and the physical environments, and those between the social and the spatial: often, the latter involves consideration of the built environment also, while spatial influences are occasionally introduced to the former. (It might be stressed that the spatial environment is viewed here as an influence on behaviour, not merely as a plane on which the other three are mapped.)

In its attempts to master the vast range of relevant behaviours, human geography has spawned a number of sub-disciplines, which focus on specialized areas: resource geography, transport geography, agricultural geography and urban geography are examples of these. Often they overlap more with other disciplines than with other sections of geography, though they share a common interest in environmental inter-relationships. Each of these fields is actively developing its own body of theory.

One of the problems of this present course of development is that it suggests a centrifugal tendency within human geography, with the production of a lot of unrelated research results. It is necessary to bring these together, to create general syntheses both at the advanced research level (Isard 1969) and as introductory texts. The latter is the aim of the present book. Ordering of the whole body of material from all of the major research areas is a mammoth task: it has recently been attempted (Haggett 1972), but the framework is as yet poorly outlined and fuller construction would be beyond the scope of the present small volume. So concentration here is on one of the main inter-relations only, the interactions of the social, spatial and built environments.

Several other texts have essayed this task, from a variety of viewpoints. Some have stressed the geometry of human spatial behaviour, focusing on the influence of the spatial environment (e.g. Haggett 1965); others have indicated the role of the latter environment as a control and constraint on a variety of human activities (e.g. Morrill 1970). The stance adopted here is to view the spatial environment as an influence on behaviour within the social environment (Johnston 1972a, 1972b), and so the material is arranged according to the main dimensions of the social environment which are relevant to the spatial approach. As outlined in the next chapter, this involves the concept of a spatial system within which aspects of modernization – here termed urbanization – occur.

The aim of this book, therefore, is to demonstrate how combinations of social and spatial causes and constraints have created environments within which societies function. These built environments then combine with the other two to channel, constrain and occasionally cause patterns and processes of modernization, in both short and long-term time scales. To achieve this aim, material has been collated from a variety of sources, forming a selected body of knowledge – representative of a much wider research literature – which can be woven together to indicate the operation of a spatial system.

This book is an introductory synthesis. Each of its chapters is a brief presentation of a great depth of material from a variety of sources: each topic could itself be the subject for a single volume. The hope is that, by putting the results of many esoteric research reports within this general contextual framework, the contribution of this branch of human geography to the total task of understanding society will be perceived. Further, in an age when 'socially relevant' research is more frequently demanded, it is hoped that this outline of the complexity of inter-relations in the spatial system will indicate the need for, and use of, such understanding in the even less tractable task of 'improving society'.

1 Spatial systems and spatial order

Societies are complex organizations. Within them, each member performs a particular role, either one which he has chosen or one to which he has been ordained. The social system is the framework in which these roles are set. It comprises several inter-related sub-systems, which set the rules and define the norms of individual and group behaviour. Thus an economic sub-system co-ordinates the production, exchange and consumption of goods and services; a cultural sub-system defines the acceptable life styles; and a political system controls much of the allocation of roles and their subsequent rewards. It is not surprising, therefore, that the main elements of Alfred Kuhn's (1966) framework for the study of society are *organizations* and *transactions*. These operate through the supporting processes of *decisions, transformations* and *communications*, all of which are embedded in a spatial matrix. Each individual and group needs an area of land on which to perform its role; the nation needs a territory. This pattern of locations comprises a spatial system, within which the social system operates. It is composed of the distribution of people and places within a prescribed area, and the transactions between these.

Within the various systems identified here, most behaviour is orderly, following sets of rules and norms. An individual's role in an economic system will require him to perform certain tasks, many of them perhaps demanding a routine reaction. There is usually a temporal dimension to such behaviour, in both frequency and specific timing, whether for an annual harvest or a daily commuting trip. In many cases there is spatial order too. The commuter usually follows the same route to and from work, the shopper usually visits the same supermarket each week, the industrialist usually collects needed inputs from the same sources and sends his products to the same markets. In such ways, the social system is spatially organized; its operation is spatially structured.

Few social systems retain the same set of characteristics for long periods of time, especially once they have crossed a critical threshold into what is generally termed the 'process of development'. A number of terms has been proposed to describe such change: *modernization* is preferred here, although it carries certain overtones that change is always for the better, following the norms set by certain 'developed' countries, most particularly the United States. Despite some differences in aims and aspirations, the process of change almost universally involves a greater complexity of social organization, and a greater volume of transactions of all types. This increasing complexity has many ramifications in the spatial system, both effect and cause. Modernization, then, is viewed here as the process by which social systems become more complex, and the focus in this book is on the spatial components of this complexity.

An important goal in the operation of social systems, and also in individual behaviour, is efficiency, which is usually measured in financial terms. This goal is a feature of spatial systems too, in their evolution and operation. Thus efficiency in movement patterns is an important influence on location decisions, on where to place a new factory or where to buy a new home, since each unit in a spatial system is an origin and destination for spatial transactions, such as the movement of raw materials to a factory or of housewives to shops from homes. There are many constraints on such behaviour, however. The results of many location decisions are often fixed, and thereby introduce inertia to the system: towns, for example, are rarely moved. Future decision-making, therefore, is often constrained by the past, which could introduce inefficiencies. Hence in studying the operation of a spatial system at any point in time, special attention must be paid to its evolution because of the great inertia produced by immense capital investment in such facilities as towns and routes.

Thus the study of spatial systems involves investigation of both the locations of functions – the *organization* component – and the interactions among these functions – the *transaction* component. Major stress in this book is on the spatial order in these organizations and transactions. Reasons for such order are outlined in the next section. Spatial systems are developed by societies, providing frameworks within which they then operate. Thus to illustrate the basic premise of spatial order, the book is structured around a simple model of societies, which is presented in a later section of this chapter.

The bases for spatial order

As already indicated, the efficiency criterion for social organization includes the development and operation of a spatial structure. This involves the efficient location of functions and organization of interactions, the flows of goods, people and messages. Locations and routes vary in the degree of efficiency they offer, because of the following flow characteristics of distance.

(1) *Overcoming distance incurs costs*   Any movement, over more than some minimal length, will involve financial expenditure, either directly in the form of fares or similar payments, or indirectly, as in the costs of shoe leather to the pedestrian. These transport costs are not often a linear function of distance. More usually, the rate of increase declines as the length of the journey increases. A major item for any movement is the fixed costs, including the provision of the infra-structure (the vehicle and the route), which have to be met. Some costs do not vary with trip length: the cost of loading a lorry is the same whether it is to travel ten miles or a hundred. For example, in 1956 it cost £2·10 per ton to move coal from Tyneside to London, of which £1·40 went on handling costs at the two termini (Manners 1964, 60). The rate of £0·70 for the journey itself suggests that the cost for a trip three times as long (£3·50) would be less than double. In fact, it may be even less, since the longer the trip, the less time will the vehicle be idle. Costs may be affected by the type of trip, also. For example, commuting trips are often relatively expensive because a large number of vehicles is used for only a short portion of each day and the traveller must pay for their depreciation while they either stand idle or are moved about nearly empty.

As an expression of this relationship between distance and travel costs, railway rates are often defined for distance zones of increasing width. Such was the case in Wisconsin in the 1950s. The rate increased every five miles in the 40–100 mile zone, every ten miles between 100 and 250, and every twenty thereafter (Alexander *et al.* 1958). Political and economic systems occasionally distort the relationship even further. Steel purchasers in the United States, for example, pay the cost of the product at the mill, plus necessary transport costs. In Britain the country is zoned, with a fixed price throughout each area. In parts of Europe, transport costs are computed from fixed basing points, irrespective of the actual origin of the steel (Warren 1965). Costs also vary by mode. Road transport is usually cheapest for short journeys; rail is cheapest for trips of medium length; shipping is the cheapest for long journeys. (Part of the cheapness of road and shipping transport is that, unlike rail, they do not have to pay directly for the upkeep of their route.) And finally, within each mode there are economies of scale of operation: the ton-kilometre cost for a 100,000 dwt oil tanker, for example, is only 38% of that for a 16,000 dwt craft (Manners 1964, 54).

Overcoming distance involves cost, therefore, for virtually every type of interaction. Cost structures vary considerably, by mode of transport, by distance travelled, and by the orientation of the transport or communications systems. The overall effect is to influence various prices within the social system. In the movement of goods from a factory to a market, for example, the cost of transport must be incorporated into the price paid by the customer. Thus the longer the journey, the higher the price. For the customer, who has only a certain amount to spend, the higher the price he has to pay because of transport costs, the less value he gets for money. For the industrialist, the closer he locates his factory to its market, the lower will be his price, which is much to his benefit in a competitive situation. (This example assumes that no other costs vary by location. The costs of assembling the inputs of fuel and of labour are all taken as invariant with location, assumptions which are more fully considered in the next chapter.) Similar arguments apply in many other situations.

It can be deduced that transport costs introduce inefficiencies into a system. The greater the amount of movement and expenditure on this, then the higher the price of goods and services will be. Assuming that income levels are fixed,therefore, as more is spent on transport so general living standards will decline. Expenditure on transport can be viewed as consumption missed, not only because of the direct costs of moving the products but also because capital will have to be invested in transport and communications infrastructure. Transport costs are, of course, not the only elements to be considered when a system is being designed, or when individual location decisions are being made. In many cases, increased transport costs may be more than correspondingly offset by declines in other costs. Presumably the goal of any social system is to operate efficiently, and thereby hold costs to acceptable levels. One of the main theses of this book is that movement costs are a significant proportion of total costs in many sectors of a society's operation. Reduction of such transport and communications costs to acceptable, if not minimum levels, is therefore proposed as a major efficiency goal in spatial system operation.

(2) *Covering distance involves time* Even in the 'shrinking world' of the 'global village', any communication between places takes time. To some people, for example peasants with three months to wait before harvesting their crops, expenditure of time may be unimportant. But in many types of spatial interaction, time costs may be at least as important as distance costs. Just as people have fixed financial budgets (occasionally varied by credit procedures) so they also have fixed time budgets. Time spent in movement may be economically wasteful. For some products, perishable foods, for example, and daily newspapers, the time element is crucial; it may, in effect, be a quality of the good or service. Again location close to the market will reduce these inefficiencies in the system. To get as much value as possible from their time budget, therefore, decision-makers may use travel time as a major

criterion determining their choice of, say, homes relative to their work places, or markets relative to their establishments.

The time taken to cover distance also affects the efficiency of a spatial system's operation in financial terms. Assuming a relatively inelastic demand for a product, the further it has to be moved the larger is the needed investment in transport facilities. One tanker may be sufficient to provide oil for Jamaica from Venezuela; several would be necessary to provide the same amount per unit of time in New Zealand, because it takes longer to reach the latter market. Furthermore, in order to insure against the possibility of even short-term dislocation of supply, New Zealand would have to hold larger reserve stocks than would Jamaica. This would involve capital lying idle, and increase the cost in the distant market.

(3) *Distance constrains available information* Interaction of the financial and time costs of covering distance produces two other important influences on the spatial system. The first of these concerns information. For example, most people are likely to know more about places near to their homes than about areas more distant from them. Their travel patterns are likely to be spatially biased, focusing on their homes, with density of contact decreasing with increasing distance. Since first-hand experience is a major source of knowledge, information will be spatially biased too. Although mass media often introduce much information in a non-spatially biased form, this information is often of little value unless reinforced by more personal knowledge. Thus, in the spread of a new drug in an American city, whereas most medical practitioners first heard of it from an impersonal source, they did not usually adopt it until after they had had subsequent discussions with other doctors (Coleman, Katz and Mendel 1966). Thus 'effective information' is likely to be spatially biased (Hägerstrand 1966) since time and money restrict the area in which people may interact and discuss such innovations.

(4) *Distance constrains opportunities* Following from the previous three premises, it can be suggested that relative locations, and hence distances, influence the opportunities perceived and the choices made. Within a given time and cost budget, each decision-maker will be seeking an efficient answer to a spatial problem, one that is satisfactory within a given set of constraints. Thus housewives searching shopping centres for a certain product, householders looking for a new home, industrialists seeking for factory sites, all tend to search within a limited area around an origin (present homes in the case of the householders, for example) and there is a high probability that their choice will not be far from their starting point.

These four ways of equating distance with various aspects of spatial behaviour suggest that the evolution and operation of a social system are very much constrained by the spatial matrix in which it is set. This is not spatial determinism, however. Each decision-maker's information field* will contain unique elements; each will react differently to the constraints, depending on aims, attitudes to time and money, and their availability. Over time, too, the influence of distance changes. Improvements in transport and communications technology produce what has been termed time–space convergence (Janelle 1968), with places coming closer together in travel time and cost, and a consequent enlargement of information fields and perceived opportunity sets. Such changes, of course, take place within an existing system, and readjustment to them is not instant. Again this produces inefficiencies in the way a social system operates in its spatial environment.

The categories of the spatial system

There are many functions in a modern social system. Many people perform more than one: as husband, father and employee perhaps, and possibly as trade union secretary or football club captain too. Some of these functions are part of the economic sub-system, others of the cultural or the political; some may overlap two or more sub-systems. All of them are in some way involved in the spatial operation of the social system. To study this multitude of functions, and the interactions which knit them into a society, they are grouped here into three major categories; structural, demographic and behavioural. These provide a three-dimensional model of the spatial system around which the rest of this book is structured.

THE ECONOMIC OR STRUCTURAL CATEGORY

Production and exchange, to enable consumption, are the bases of a society's operations and just as each individual person has a specific role, or roles, in this organization, so too has each location. Large tracts of land are given over to the production of food or other raw materials, whereas in a few places many people congregate, to exchange the products of the land, to process these and exchange them again, to provide the necessary organization for this trade, and to facilitate the provision of other services which are part of the local standard of living. Where are these concentrations? How large are they? Why do they develop? Which functions locate in which place? And what are the flows between the places?

Social organization involves functional differentiation, therefore, which in turn produces a spatial division of labour. In Kenya, for example, most of the control functions – political, social and economic – are concentrated in Nairobi; many of the other economic roles, plus less important (spatially more restricted) control functions, are performed in a few other towns, mostly either close to Nairobi or arrayed along the country's main transport corridor leading to Mombasa. Other rural areas are articulated into this spatial system; they perform more specific roles, usually specializing in the production of only a few commodities. More remote areas are politically controlled from Nairobi, but play only very minor roles in the economic system; the most remote are not yet effectively integrated into the Kenyan social and economic systems (Soja 1968).

Exchange is an integral part of the operation of such spatially separate areas, so that economic systems are structured around their towns, roads and railways. Indeed, these three, plus the financial institutions which invest in them, are the catalysts for expansion of a system into as yet unintegrated territory (Gould 1970a). As this areal expansion of modernization proceeds, spatial inequalities are usual. Often two economic systems exist in tandem, the traditional and the modern. With time, the former may integrate with the latter, perhaps modifying traditional ways to adopt to new social norms, or alternatively completely assimilating the novel behaviours introduced by the modernization process. In some cases, the traditional elements may be exterminated (as with Tasmania's Aborigines) or forced back into the least attractive parts of the spatial system (as with the New Zealand Maori and the Aborigines of mainland Australia). Even full integration of all areas into the one organizational structure does not remove the spatial inequalities or reduce the concentrations, however. As has been demonstrated for Canada (Ray 1969), a spatial morphology similar to that described above for Kenya is still typical.

Modernization produces continuing division of labour, therefore, a division which has spatial form. In any social system, international, national, regional or local, each function must be in a place, and must interact with a range of other functions. This economic category, then, defines the first dimension of the spatial system.

THE DEMOGRAPHIC CATEGORY

As a corollary of the concentration of functions suggested in the preceding section, modernization implies a process of population concentration. Two factors associated with the efficiency criterion encourage this process. First, efficiency is advanced by the achievement of what are termed internal economies of scale. Large organizations are usually more efficient than small ones, at the same task, especially if this involves production of goods rather than services. This is because large firms can make better and fuller use of their capital, equipment and labour, so that their production costs per unit are relatively low. Specialization of role within the larger organization (a more complex division of labour) is a major contributor to these economies. (It is, of course, also possible for a firm to become too large and unwieldy so that with further growth diseconomies set in.) The second factor concerns external economics of scale. Each organization, firm and household interacts with many other such units, in the exchange of goods and services. Such interactions are generally much easier and cheaper if the units involved are in the same population concentration.

There is, then, a general relationship between modernization and population concentration, which together produce *urbanization*, which might be defined as a major spatial concomitant of the processes of social change. Indeed, initial concentration of people is probably necessary for generating modernization processes. The relative growth of the main population clusters may be due to faster rates of natural increase there, but usually, as detailed later, its main component is migration. Thus in Kenya, the main centres of modernization are also the main centres of tribal mixing as a result of such migration. In Nairobi and Mombasa, the main urban, organizational nodes, the two largest tribal groups account for only about one-third of the population; in the least modernized areas, which have no large nodes of population concentration, a single tribe may alone account for 99% of the residents (Soja 1968). So the city, the keystone of the economic dimension of modernization, also becomes the 'melting pot' of people drawn from various backgrounds and cultures.

THE BEHAVIOURAL CATEGORY

So far, modernization has been discussed with reference to both a spatial division of labour and a concomitant process of concentration in the distribution of population. It is also associated with changes in life style, or patterns of human behaviour. As modernization proceeds, so ways of life tend to change. The larger the place one lives in and the larger the organization one works in, the more restricted one's role in society will become. More interactions will become necessary for living, an increasing proportion of these interactions will be at an impersonal level, and the more segmented life will become. On the other hand, larger places generally offer a wider choice of occupations, of jobs (within each occupation) especially for women, and of ways of spending both time and money.

The relationships between many of these behavioural changes were associated with modernization/urbanization in a classic essay, 'Urbanism as a Way of Life', by Louis Wirth (1938). The independent variables – the 'causes' – were postulated as the population size, density and heterogeneity of the city; the results were a growth in impersonal relationships and in the dominance of large organizations. (A full appraisal of Wirth's ideas is given by Morris 1968.) Thus as city growth proceeds, Wirth suggests that the kinship networks of smaller-scale societies are broken down and individualism, usually based on the nuclear family, increases. That some such changes occur is generally accepted by later scholars, although detailed research in many locales has indicated that the breakdown of 'community life' is not necessarily a produce of modernization. Indeed, it is suggested that certain behavioural changes, particularly in attitudes and in family structures, are a prerequisite for, not consequence of, modernization (Goode 1963).

This last point indicates that although it is possible to isolate the three categories described here – structural, demographic and behavioural – it must be realized that these are simplifying reality in order to assist in understanding it (which is the classic purpose of a model). The three categories represent dimensions of social processes, operating within a spatial environment, that are complexly interrelated with each other. A change in one will initiate further changes in the others, and probably, through feedback, in itself too. The growth of large urban industries causes greater job specialization, to which labour may react by the formation of trade unions (which themselves create more jobs and further division of labour). These jobs will normally be located in an urban place, to which new residents will be drawn. In turn, these make new demands on the system, perhaps initiating further job creation, increasing the cultural variety of the place, and attracting more people there. And so the process continues.

Spatial scale and the present book

The general thesis of this book is that full understanding of society requires study of the spatial system in which its economic, political and cultural sub-systems operate. Societies create spatial systems, by placing different economic, political and cultural functions in separate places, and then operating within these frameworks, while constantly modifying them to meet changing requirements. This behaviour, both in the operation of a system at any one time period and in the changes to its morphology, consists in a large part of ordered reactions to the constraints of distance.

There are many ways in which one could organize a book on this theme. One could, as others have done before (e.g. Haggett 1965), focus on the spatial constraints and morphologies; or propose the most efficient ways in which space might be organized (Scott 1971, introduces some of the relevant literature). The focus here is rather on the description of spatial systems as they exist, within the context of their relevant social systems. This can involve three foci (Golledge 1970).

(1) Description of the spatial structures at any one time, which would emphasize the differences between the component parts, and suggest reasons for these. Such an approach could be extended by studying several temporal cross-sections, tracing the changes between them (Berry 1964a), and, again, suggesting possible causes.

(2) Study of the system's operation at any one time, of the inter-actions between and within its component parts. Such investigation would necessarily involve the first approach also, since the interactions could not be studied apart from their origins and destinations. To-gether, these first two approaches would provide a full description, though not necessarily an explanation, of a chosen spatial system.

(3) Study of behaviour in the system, emphasizing the mechanisms of change (not just the processes) and the nature of relevant location decision-making. Such a focus requires a study of the bases of human behaviour; it would hopefully uncover the ways in which both distance and the spatial structure acted as constraints on future action.

Since research representing the third approach is relatively poorly developed (see Pred 1967, 1968), the main attention here is on the spatial system, its processes of change, and its operation. It has already been stressed that past decisions strongly constrain those which follow, so it is accepted that knowledge of a system's evolution is necessary for both description and tentative explanation of its present and future form. Hence an evolutionary perspective is adopted. Ex-planation of the nature of spatial systems is not being offered there-fore. The book presents a description of patterns and interactions along with hypotheses of their development and operation, thereby accounting for the spatial order (or lack of it). Testing many of these hypotheses remains the task of future research.

SPATIAL SCALE

Finally, there is the question of spatial scale, the size and nature of the areas to which the approaches and principles outlined in this chapter will be applied. In fact, they are relevant at all scales, from the largest to the smallest inhabited area, from the whole earth to an individual room. Geographers have, in fact, tended to ignore the extremes of this continuum. The study of small areas, usually parts of buildings, has been left to other social scientists (see Michelson 1970); only econo-mists pay much attention to the world system. Here the latter has been included, since it offers clear advantages for studying a spatial system on a broad canvas, although there is a paucity of material on which to draw. Two other scales have been chosen, each meeting the criterion of providing a relatively closed, or self-sufficient, system, the majority of whose interactions remain within its boundaries. These are the nation and the urban place (Berry 1967a).

Each of the remaining three chapters considers one of these systems; the national, the individual and the international. No set chapter structure has been imposed. The format and allotted space depend on the availability of material, hence the relatively speculative nature of the final chapter. In all three, the structural dimension of the three-category model dominates, both because it has been most widely studied and because it is probably the key to the whole complexity which we hope to reduce to spatial order.

2 The national system

In the operation of a national system, economies are obtained through the concentration of activities and people into only a few places; the greatest economies are in the scale of operation this allows and in the time and money expended on communications and interchange (Mills 1972). Although efficiency is an aim, however, there is little likelihood that the distribution of cities in a system, or their locations, sizes and functions, are optimal (Richardson 1972). Other goals such as equity of income distribution, may also be aspired to, and these may conflict with the goal of efficiency (Alonso 1969), while the cumulative inertia of past location decisions and invested capital is usually sufficient to ensure a 'non-optimal' pattern.

Many approaches could be followed in a chapter on national systems. That chosen here is concerned with the ways in which systems develop and become interlocked, and how they operate in terms of the movement of goods, services, people and ideas. The first sections, then, focus on the structural category (p. 9), on the organization of a national space-economy through the modernization process. The later sections, dealing with the articulation of the system in economic terms, lead into a discussion of a similar theme with regard to the flows of people and ideas.

The space economy

The question of how a space economy develops is often answered by invoking the principle of comparative advantage. This states that each place (region or town) obtains the functions for which it is best suited in the system's context, based on the relative merits of the locally available factors of production (land, labour, capital and enterprise). In the system, many places may be able to accommodate similar tasks. Some of these tasks may be duplicated, since each place will have a spatial monopoly or quasi-monopoly over a market; for others, there may be competition to see which place will be allotted which package of functions.

No location in any one system is likely to have the absolute advantage for the production of all the required goods and services, since some resources are not widely scattered and some uses require large areas of land. Thus, various functions will be scattered over a national space. (The exception to this is the self-contained, local economy, almost certainly at a very low level of economic development.) But it is dubious whether in fact the distribution of functions emerges through a competitive process, until an equilibrium of comparative advantage is reached, since development is a continuous process. In some, relatively simple functions, such as the provision of central services (as described below), a form of the competitive process may operate, but with many functions the comparative advantage has probably been imposed on an area, wholly or in part, by a socio-political system, and maintained by that system plus the inertia of an existing pattern. Hence the descriptive sketch offered here can be no more than a general statement of how spatial systems appear to have emerged.

EMERGENCE OF AN URBAN SYSTEM: CENTRAL PLACE
DYNAMICS

Obtaining food dominates life in relatively underdeveloped, self-contained, usually spatially-circumscribed societies. Emergence from this situation requires increasing productivity, which can usually only be attained through job specialization and trade. Initiation of trade is probably imposed by some external power, sometimes of a religio-military nature (Wheatley 1967), but once established this may encourage further development and trade (Jacobs 1971) while movement along trade routes may attract other groups to join the growing spatial division of labour (Pirenne 1925; Hodder 1965). Initially, such exchange of products probably developed at junctions between various physical environments, allowing the advantages of the varying resource endowments to be exploited.

The emergence of towns thus dates from the initiation of trade. Their main comparative advantage has always been accessibility to the trading groups, which advantage may be 'naturally' endowed in the landscape because of site and situation, or a human creation as a node on a system of man-made routes. Once a trading post was established, people habitually visited it to exchange products, more routes focused on it, and a permanent settlement developed (though there were exceptions to the latter: Hodder 1961). Most such settlement systems developed under primitive transport technologies and reflected the limited spatial horizons of their inhabitants. As a result the settlements were only small, articulating the exchange of products generated by the local population. (Much work has been done recently on such trade systems in Africa: see Smith 1972.) Usually, however, wider spatial horizons existed for at least a small minority of the population, either immigrants from another system seeking trade (Vance 1970) or local residents seeking to extend their personal influence. Because of these, hierarchical systems of markets developed, comprising nested cells of local and more extensive systems of interchange. The British system for example, comprised both the local market towns, the foci for areas with radii of perhaps six miles, plus a set of fair centres, at which traders (and others, such as assize judges) from much wider areas met at fixed dates (Dickinson 1934). And above these was at least one other larger centre, the national capital. Through such hierarchical systems, the products of relatively isolated rural areas were distributed widely. The size of a town was largely a function of its position in the system; the larger the area it served, the larger its permanent population, and the larger the transient population which visited it on set dates.

The development of trade, whether symbiotic in terms of the exchange of one product for another, or parasitic when it represents the extraction of tribute by an elite, parallels the emergence of towns (each become cause and effect for the other). Initially, the trading function is a temporary one, perhaps on only one day each week. Eventually, however, it becomes a permanent role, as traders open full-time establishments in each place. This occurs when trade potential increases, usually through greater productivity of the town's hinterland or sphere of influence from which patronage is drawn. (An additional, rather than alternative, process leads to the selection of only some of the more accessible centres as permanent towns, because increasing personal mobility allows traders in places with that comparative advantage to extend their hinterlands and capture those of less-favoured places.)

This brief sketch has outlined the development of a central place system, a settlement pattern whose nodes function as collection and distribution points. Because the various central services vary in their thresholds, or levels of demands, a hierarchical system of centres is usual. Some services are used frequently by most people – such as food stores – so each establishment's* operation can be supported by a small hinterland and there are many such establishments distributed through the space economy. Other services, the selling of clothes, for example, are less frequently demanded. There are fewer establishments, therefore, located in a smaller number of centres, to which the average customer must travel further than he does for his more frequent wants. In such a situation, there would be two types of towns, one selling only food, the other both food and clothes. (Towns with clothing but not food shops are not expected; see p. 18.) Ideally, the number of types of towns would equal the number of types of establishments. In practice, however, many types of establishments have similar thresholds, so groups of central services distinguish types of towns in the hierarchical system, rather than any individual functions.*

There is a vast literature on the service functions of urban places, on what has become known as central place theory (Berry et al. 1966). Most of it stems from an initial statement by Christaller (1933) who, assuming the omnipotence of the efficiency criterion for a homogeneous population inhabiting an infinite, homogeneous plain, demonstrated that the hierarchical system of central places would be spatially arranged with a series of nested hexagonal hinterlands. In fact he suggested three possible alternatives, one based on the efficiency criterion alone, one assuming also a grid system of routeways, and the third assuming spatial independence between the various hinterlands – fig. 2.1: see Berry 1967b. The first of these is usually taken as the model against which reality should be compared.

The development of central place systems has been less intensely studied than their existence and operation at any one time, mainly because of the lack of data. Skinner (1964–5) has reported a detailed investigation of a system's emergence, comprising probably six types (orders) of town, in mainland China. The basic element in the evolving pattern was the standard market town, providing an average of about eighteen villages with both itinerant and permanent establishments (produce traders characterized the first group, eating and drinking places the second). The larger, less frequent centres performed two types of function.* First, they marketed only the goods and services purchased by an elite of the total population, hence the need for a large hinterland; secondly, they facilitated the diffusion of more exotic goods to the standard market towns, and provided a reciprocal articulation of the flow of certain items of peasants' production away from their local area. (That most people made little use of the larger

---

fig. 2.1 *The size and spacing of central places, plus their hinterlands and connecting transport routes, according to the variants of Christaller's theory. A shows the most generally used, market principle, which is the most efficient in terms of the number of centres, but not in the transport network needed. At left is the hierarchy of places and the needed routes to connect each place to its nearest higher order centre; at right is the hierarchy and the hinterlands: note that each smaller centre is at the intersection of the hinterlands of three centres of the next largest order. B shows the transport principle, which minimizes the route mileage necessary. Each centre is on the border of only two higher order hinterlands in this scheme. Finally, C shows the administrative principle, in which each hinterland is discrete and lower order centres nest within the hinterland of only one larger centre. Note the resulting complexity of the transport net.*

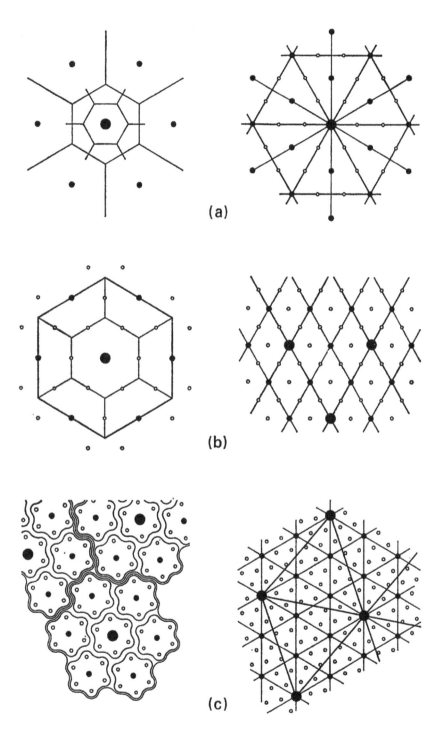

(a)

(b)

(c)

Hierarchy and hinterlands   Hierarchy and routes

centres is shown by the example of a family Skinner stayed with. During three months they visited the local standard market town 46 times, but went on only three occasions to the next largest centre, which was less than twice the distance further away.) Thus a hierarchical spatial structure emerged, with the local circulation system meeting most needs, but with higher levels of organization articulating the long-distance trading within the total space economy.

EMERGENCE OF AN URBAN SYSTEM: INDUSTRIAL PATTERNS

The preceding paragraphs have described how a system of towns develops to serve as the foci for inter and intra-regional trade (in which the region is the urban hinterland). In the ideal case, in which the population of a town is a function of its importance in the system, the size distribution of these settlements should be stepped in form (see fig. 2.3) comprising groups of places of similar size, and their spatial distribution should display a nested hexagonal form, with each town at a certain level in the system serving a set number of towns at the next lowest level. Such perfection in spatial organization is never achieved, partly because the real world is not as simple as suggested in the assumptions of Christaller's model – there is no homogeneous plain on which all decision-makers act, successfully, to gain complete economic efficiency. In part, it is because of the model's failure to deal with dynamic processes, including locational inertia. It is also true that the theory is only a partial representation of urban functions, since it ignores the industrial role of urban settlements, focusing only on tertiary activities.

Some industries are distributed through a space economy according to central place principles. Their materials are usually either ubiquitous or cheap to transport (at least relative to the total costs of the product); their products are widely demanded, frequently bulky, and may require virtually instant delivery to a local market, often because of perishability. Hence firms such as bakeries, breweries and newspaper printers (especially printers of evening papers in many countries) are often widely scattered through a system, closely reflecting the distribution of population. Such firms vary in their scale economies, however, so that the size of towns in which they are located may vary too. In general, the lower the transport costs relative to total costs for the industries, the larger will be their local, possibly monopolistic, markets and the fewer the number of towns in which they are sited. A hierarchy of manufacturing industries distributed through a hierarchy of towns is thus a possible feature of the spatial system.

Most industries do not enjoy local monopolies and must compete for the trade of all, or a large part, of the system. In studying their location patterns, it is necessary to classify industries. The scheme used here (following on Duncan *et al.* 1960) is based on the general nature of factory inputs and outputs rather than any specific products. Two basic types are identified, processing and fabricating. The latter produces a final product for the market, whereas the outputs of the former are only partly-finished goods needing further processing. Each type can be further sub-divided according to the nature of the inputs, whether these are raw materials or partly-processed items. Some inputs are very far removed from their raw state. In general, the costs of materials in the total price structure of a product tends to decrease relative to other costs, notably labour, the further along the processing chain a factory is located. In New Zealand during 1968–9, for example, wages and salaries formed 18%, 26% and 46% of the total costs of, respectively, metal founding, sheetmetal working and electro-plating/metal polishing.

The greater an industry's dependence on raw materials, especially bulky raw materials, the more specific its available locations are likely to be. In most countries, raw-material-oriented locations form the bases for present industrial distributions, since material-dominated processes, as in the iron and steel industry, have generally been among the first established. In addition, at early stages of economic development material costs have usually been high relative to those for labour. Processing a local material, agricultural, horticultural or silvicultural, usually initiated manufacturing in a town, which may have already been a central place.

Not all towns which obtain industries grow to metropolitan status. Which do and how? Thompson (1965) has suggested a model in which the process of urban growth is divided into five stages.

(1) *The stage of export specialization*, when the town is dominated by a single industry, perhaps by a single firm. At such a stage prosperity may be ephemeral, especially if the industry is based on an exhaustible resource and all of its production is exported to other parts of the system.

(2) *The stage of export complex*, in which the town's economic future becomes more assured and other industries, associated with the original one, are attracted there to form a limited industrial complex. In many cases, these new industries may process the outputs of the initial firm, as in the chemical industry at Ashtabula, Ohio (Hunker and Wright 1963) and in steel-making (Isard and Kuenne 1953). In others, the new plants may be opened to provide inputs for the original factory. This latter process is termed backward linkage; the former is known as forward linkage. In modernizing countries, it is often hoped that spontaneous backward linkages will follow some initial industrial development, as happened in Japan where components for the assembly of bicycles were initially imported but were later produced locally (Jacobs 1971).

(3) *The stage of economic maturity.* Central functions may be attracted to a town in one of the first two stages, and perhaps even a few small industries to serve the local population also. The local market will probably be too small to attract many, however, and it will be reliant on other places for much of its needs. If its basic industries expand, the town may become large enough to attract new fabricating industries to serve the local market only. These will be industries in which raw material prices are of little consequence in their total cost structures. The process is termed *import replacement*. It initiates what is termed an *urban multiplier*, in which growth creates more growth (Pred 1965a). A new industry is attracted to the town, making more work available there; its inputs may include outputs of other local factories, generating more work there, too; its employees will spend their incomes in the town's shops, creating more job opportunities in them and perhaps in other local factories. In combination, all these processes create a larger market in the town, maybe sufficient to attract further new industries.

This import replacement multiplier is cumulative, therefore, in its role of attracting new industries. But it only begins to be effective when the town reaches a considerable size, which the consensus of opinion places between populations of 100,000 and 500,000. Beyond that critical threshold, the attraction of more and more fabricating industries for which market size is a major locational factor (and these become more common as technology progresses and processing chains become longer) means that a place becomes more self-sufficient in meeting local demand. This is shown in the positive correlation between urban size and the ratio of basic to non-basic employees there (Blechynden 1964); basic employees produce for markets outside the town itself and non-basic employees for the local market. The larger the place, in general the smaller the number of 'exporters' relative to the total number of workers.

(4) *The stage of regional metropolis,* during which a new growth process, *export diversification*, comes into play. In this, industries grow to serve external rather than internal markets. Such industries may be entirely new ones, 'invented' in the town itself or located there because of various combinations of advantages. Usually, however, they will be extensions of existing industries. Factories producing only for the local market may begin selling beyond it, perhaps because their larger size relative to factories in other towns allows them to produce at a cheaper price and hence compete over wider hinterlands. Alternatively local residents may develop factories to process the byproducts of existing factories. The larger the place, the greater the concentration of such initiative, and of the resources, such as capital, to foster it. In the U.S.A. there has been a concentration into the

largest cities of patent-granting, although this concentration has declined as more regional metropoli emerge. In uncertain market situations, too, the larger places probably offer the greater chances of success, or buttresses against failure (Webber 1972). Certain types of city also offer better environments for export diversification, perhaps because of the generally smaller size of firms there: Birmingham, England, has been favourably compared with Manchester, for example (Jacobs 1971), Boston with Detroit (Thompson 1965), and New York with Pittsburgh (Chinitz 1965).

New industries generated through export diversification also act as further primers for the process of import replacement, since they too will have reciprocal effects on production in many other factories. (The growth created may be in jobs or, as productivity improves, in real incomes.) A large city in this stage, therefore, is likely to grow at an exponential rate, much faster than one in the economic maturation stage: indeed, Jacobs (1971) has maintained that a city differs from a town in its participation in both processes; a town is only engaged in import replacement.

(5) *The stage of technical and professional virtuosity* The nature of this stage is not altogether clear, but its main features are national or even international domination in production of some goods or services. As pointed out below, a town need not pass through the whole process in order to attain this stage.

According to Thompson's model, towns begin with a single function, and growth involves the expansion of the economy through linkages to other plants and to the local market. Further growth may be generated by the addition of new functions providing for an external market. It is possible that a town's initial role could be as a central place, but few such places would have large and productive enough hinterlands to propel them over the critical threshold into the maturity stage. More usually, then, it is a processing industry which forms the basis of growth, perhaps in combination with a central place function.

What type of settlement pattern is generated by such a process? It is feasible that all towns in a system could reach the maturity stage, but very unlikely, since most would retain comparative advantage in at least one function, thus placing them all in the fifth stage. Most likely, only a few places will proceed into the third and fourth stages; the initial advantages they possess will ensure that they benefit most from scale and other economies. Since in any system, such initial advantage is almost certain, in the light of uneven population distributions, productivity of primary industry, or just the location of government and private decision-making, then some towns will be able to outgrow others. Where industrialization is grafted on to a

c

central place system, then, it is the largest centres which are likely to prosper, especially with regard to the twin processes of import replacement and export diversification. Only towns which grow because they are located near to valuable fixed resources are likely to produce any marked alteration of the rank ordering of centres by size.

This importance of size is shown by an analysis of the locations of processing and fabricating industries within the United States (Duncan *et al.* 1960). The former are fairly widely distributed among towns of all sizes; the latter are much more concentrated into the larger places. But size was not the only determinant of industrial location patterns, another was relative location of a settlement. This was measured, following Harris's (1954) lead, by computing the population potential for each point with the formula:

$$Vi = \Sigma(Pj/dij)$$

where
- Pj = the population of the jth centre
- dij = the distance between i and j
- Vi = the potential at i

and summation is over all places $j = 1 \ldots n$

The larger the value of Vi, the closer the place is to the entire national market. This greater accessibility is positively related to the distribution of fabricating industries (Duncan *et al.* 1960), the rate of industrial growth (Harris 1954), and the distribution of high value-added industries, those for which labour costs were over 70% of total price (Pred 1956b).

This last point substantiates the earlier argument that towns may enter the stage of professional and technical virtuosity without first passing through the other four. If town size alone were related to the distribution of fabricating, Thompson's model as outlined would be valid. Since relative location is a further independent variable accounting for the distribution, however, this indicates that small towns near to large markets can attract important fabricating industries. Small towns in the hinterland of New York, or Toronto, or London, for example, may thus be able to attract a specialized industry serving a wide, large market.

It is not only in the distribution of secondary industries that these general principles apply. Tertiary employment largely follows the distribution of population, indicating the relevance of the size variable. But smaller centres close to large agglomerations may be able to attract important service industries, especially quarternary, or administrative, industries (Goodwin 1965); the concentration of insurance companies in Hartford, Connecticut, is a good example of this. As Chisholm (1971) and others have pointed out, this importance of accessibility may not be a function of movement costs, especially in a small, densely-peopled country such as Great Britain where transport costs form a small part of most delivered prices. Such costs may, in any case, be counterbalanced by lower land and labour costs; rather, as a survey of Tayside manufacturers showed (Begg 1972), it is remoteness from personal contacts and ancillary services in the industrial field, plus the time involved in moving products, which are the major disadvantages of peripheral locations.

Finally, the accessibility factor also influences the location of primary industries. As long ago as 1826, von Thünen studied this spatial pattern (Chisholm 1962; Hall 1966). His results suggested a zonation of horticultural, agricultural and silvicultural land uses around a market, and this pattern has been adopted as a model to be tested in a legion of empirical studies. The model is based on the rent which

TABLE 2.1   *The spatial structure of agriculture: Oklahoma, 1960*

Miles from nearest dominant urban place	Farm people/ 100 acres farmland	Average farm size (acres)	Cows milked/ 100 acres farmland	Productive man work units/100 acres farmland
0–9	4·3	141	3·8	67
10–19	2·8	184	2·9	54
20–29	2·0	222	2·2	43
30–39	2·0	221	2·0	42
40–49	2·0	216	2·1	41
50–59	2·1	198	2·2	45
60–69	2·2	196	2·3	48
70–79	2·0	194	2·4	49
80–89	1·9	236	2·1	41
90–99	2·2	224	2·1	44

Source: Tarver, Turner and Gurley (1966)

can be obtained from any piece of land. Certain products cost more to transport to the market than others, relative to their total costs. Others need to get to their market quickly, usually because of their perishability. Such uses therefore bid more for the more accessible land than do those for which transport costs or time are less important. They will not, of course, pay more for their land than they can afford: the costs of an accessible location (production and transport costs plus rent) must not exceed the benefits (the price obtained at the market).

This competition for land, with producers balancing transport costs against rent, produces a zonal pattern of uses. Often this involves a greater intensity of use close to the market, as shown for the dominantly crop-producing state of Oklahoma in table 2.1. But this relationship may not always hold. In von Thünen's scheme, the second zone out from the market was occupied by forests (horticulture occupied the first). This relatively extensive use of land close to the market was located there because of the large, fairly inelastic demand for timber—for fuel as well as building—in the early nineteenth century, and the high costs of moving this bulky product.

Many studies have indicated the general value of this zonal model. Not only is it relevant to the location of land uses relative to their market; it also can be used to account for their location with regard to the producer's home. The further a farmer's field is from his farmhouse, the less frequently will he visit it. As a result, there is a zonation of uses, and often of intensity of use, with distance from the house, a

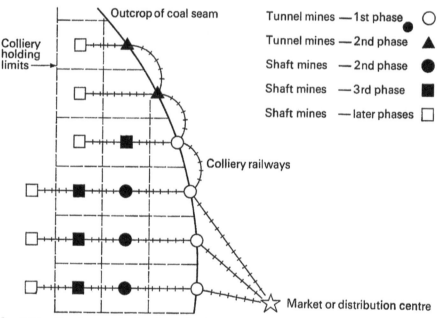

fig. 2.2   *Development of a hypothetical coalfield away from a market centre.*
Source: Wilson (1967).

pattern which is most noticeable either on large holdings (Chisholm 1962, 51) or where farms are nucleated into agricultural villages (Blaikie 1971).

With other primary industries, too, zonal patterns of exploitation may be observed. On the New South Wales coalfields, for example, initial mining was close to the market/port at Newcastle (fig. 2.2), and with time moved further into the hinterland. Forest exploitation may display the same spatial order over time.

This discussion of certain spatial aspects of all sectors of an economy has been focused on urban growth patterns. In effect, the same argument applies to regional as well as urban development processes. Since development is closely associated with population concentration and a basic function of towns is to organize their regional hinterlands, this focus on towns does not distort the essential elements of the described process.

Urban size distributions

In the preceding two sections, various aspects of urban growth have been described; the present section deals with the end-result. Do all systems have a similar distribution of city sizes—a few large, several of intermediate sizes, and a lot of small ones, for example? And if they vary, are there any underlying reasons for the differences?

HIERARCHIES AND THE RANK-SIZE RULE

The theoretical discussion of the emergence of a central place system suggested that it should result in a stepped distribution of urban sizes because of a similar stepped distribution of thresholds for central services (fig. 2.3A). Many studies, however, have suggested rather that there is a continuum of settlement sizes, usually termed Zipf's rank-size rule, which forms a straight line on double-logarithmic graph paper since its most basic form is

$$Pr = Pi/r$$

where Pr = the population of the relevant city
Pi = the population of the largest city in the system
r = the position of the relevant city is a rank-ordering of all cities in the system from the largest to the smallest.

According to this formula, the ratios of size and rank-ordering are constant for any pair of cities. A city which is half the size of another, therefore, will have a rank-order position double that of the latter (fig. 2.3B), though in many systems r must be weighted by a constant to fit the data set.

These two distributions would appear to conflict. In fact often they do not, since the rank-size rule may be fitted to grouped data (the number of towns in a certain size range), for which a hierarchical distribution may fit equally well if the population of each individual place was studied. Nevertheless, there are cases where a rank-size continuum can be observed, as for the USA, perhaps as a result of variations in the environments in which the central place principles operate. Any one system may comprise several different physical, social or economic environments supporting different population densities or settlement patterns (Johnston 1966a). As fig. 2.3C suggests, each of these environments may contain a hierarchy, but comprising settlements of different sizes (Brush and Bracey 1955); when these are amalgamated, they form a continuum. Alternatively, Lösch

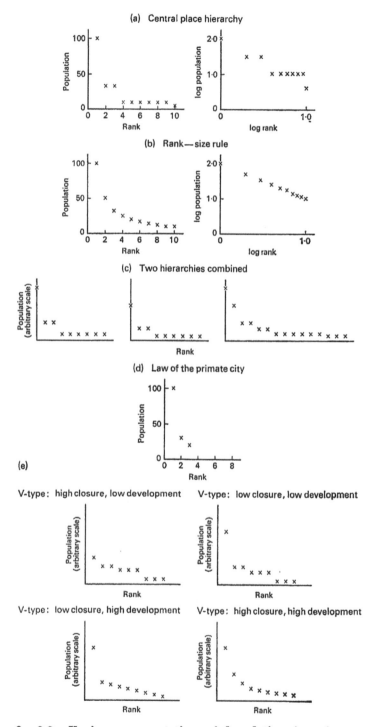

fig. 2.3 *Various representations of the relative sizes of places in an urban system.*

(1954) has suggested that establishments serving different market areas do not necessarily cluster, as is the case with Christaller's scheme in which any high order centre contains all of the functions of a lower order centre. If such nesting does not occur, a continuum of settlement sizes is one possible result.

In his formulation of the rank-size rule, Zipf (1949) suggested that it resulted from a balancing of the forces of concentration and dispersion, in an argument somewhat akin to Christaller's. But the rank-size rule is often viewed as an empirical regularity for which no rationale is known. It has been advanced that it is the most probable result of a random population distribution, but Berry (1970a) has refuted this. Instead, he has demonstrated that if one starts with a random distribution of population among a given number of settlements, and if the same growth rate applies to all members of each town-size group, then a rank-size distribution will ensue. Unfortunately, it is well known that urban growth rates vary widely. More recently, Pederson (1970) has hypothesized that diffusion processes may act to transform a hierarchical system into one with a rank-size pattern. The model assumes: (a) one large place (A); (b) two places (B and C), smaller than the first, and of the same size as each other, but one (B) spatially much closer to the largest than is the other (C); (c) innovations originating in the largest place; and (d) the spread of an innovation is spatially biased, with early adoption close to the origin (see also p. 38). Thus, of the two cities in the second size group, that closest to the largest will receive all innovations first. This will give it initial advantage in the growth process through the import replacement sequence (p. 22) making it larger than its rival of the same initial size. For example, the largest place (A) may be the site of the first university, in 1900. Place (B) gets one in 1920, the extra population that this generates making it larger than place (C), whose university opens in 1930. So, for a time at least, place (B) is larger than (C) and the urban size distribution more closely approximates the rank-size rule than the central place hierarchy. And if innovations are forever being diffused from (A), place (B) will maintain its size advantage over (C). This process can operate throughout the central place hierarchy, and not only at its upper levels.

Perhaps a final approach to this problem is to suggest that the introduction of industry to a central place system produces certain distortions to the stepped pattern. Because of local resources, some small centres may gain industries and move into the export complex stage, thus making them slightly larger than their former peers. This may occur at all levels, producing size variability which will retain some of its hierarchical form because of the relationship between urban size and import replacement: hence the rank-size pattern.

THE LAW OF THE PRIMATE CITY

In many countries, neither of the above city-size distributions applies, but the country is dominated by a single centre. This was remarked on by Jefferson (1939), who pointed out that the average percentage ratio of the populations of a system's three largest cities was 100–30–20, though there were very wide variations from this. (How he obtained this average is not clear.) Usually, however, primacy is applied only to the dominance of the largest place, producing the size distribution of fig. 2.3D.

Which countries have primate distributions and which have rank-size/central place hierarchies? Berry (1961) suggested that the former pattern was typical of underdeveloped countries, and the latter of developed, but this hypothesis did not fit reality. Linsky (1965) suggested that large size prohibited primacy and that among small countries, low average incomes, high population growth rates, a colonial history, export orientation, and agricultural domination were all conducive to primacy. However, as many small countries did not have a primate pattern, he could not indicate the prime causes of the phenomenon. Nor could he, like Berry, account for countries such as Australia which have neither primate nor rank-size pattern.

The best account of 'why primacy' for certain situations would appear to be the 'colonial' (Rose 1966) or 'gateway' (Burghardt 1971) model. Many countries were settled from overseas through ports in which administration, commerce, and incipient industry for the extensively farmed hinterlands were all highly concentrated. (These 'ports' may have been inland, like Cincinnati or Minneapolis.) Because of low densities, few central places developed inland, certainly very few of any size, especially if the economy was externally oriented. The port centres thus dominated the system and possessed immense initial advantage for any further development, which they could orient in their own direction, as with Toronto's investment in Ontario's railroad system (Spelt 1955). This primacy may eventually be challenged if massive industrialization in the hinterland is possible, otherwise no other centre may even reach the mature stage. New York's primacy was challenged by Chicago, for example, and Cincinnati's by Indianapolis, Dayton and Louisville. Minneapolis's primacy was never challenged by any rising new industrial centre, however, and it, with St Paul, remains by far the largest centre in the northern high plains. In Australia, and several other countries with neither rank-size nor primate distributions, the observed pattern results from the amalgamation of several primate systems of different size, one per state.

Vance (1970) has presented a generalized cartographic model of an urban development process (fig. 2.4), in which the central place pattern

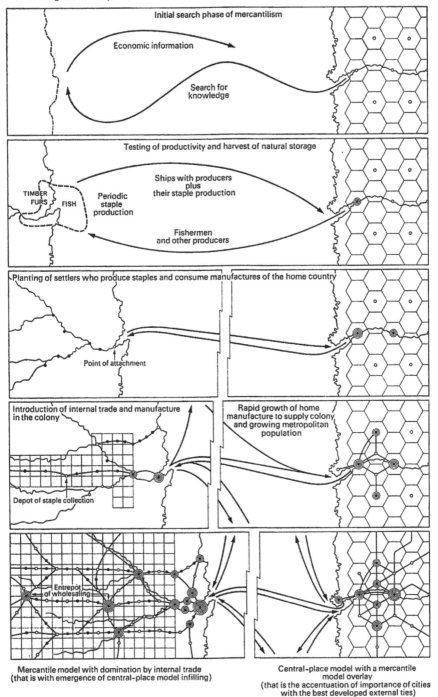

Based on exogenic forces introducing basic structure

Based on 'agriculturalism' with endogenic sorting-and-ordering to begin with

Initial search phase of mercantilism

Economic information

Search for knowledge

Testing of productivity and harvest of natural storage

TIMBER FURS   FISH

Periodic staple production

Ships with producers plus their staple production

Fishermen and other producers

Planting of settlers who produce staples and consume manufactures of the home country

Point of attachment

Introduction of internal trade and manufacture in the colony

Depot of staple collection

Rapid growth of home manufacture to supply colony and growing metropolitan population

Entrepôt of wholesaling

Mercantile model with domination by internal trade (that is with emergence of central-place model infilling)

Central-place model with a mercantile model overlay (that is the accentuation of importance of cities with the best developed external ties)

fig. 2.4 *Development of central place systems in a colony* (left) *and its 'motherland'* (right). *Source: Vance (1970).*

of the 'mother country' is eventually matched by that of the colony. (Though, of course, many colonies never reach the final stage and retain the primate pattern. Vance also argues that similar external forces have transformed the central place patterns of many 'mother countries'.) This model suggests that rank-size/central place hierarchy and primate distributions are not mutually exclusive. Instead, countries may have an amalgam of the two. Vapnarsky (1969) has related the degree of primacy to the degree of a system's 'open-ness', its dependence on external trade. Similarly, the development of a rank-size pattern is positively associated with the degree of industrialization. Four types of system are thus possible. With a two-way classification according to open-ness and internal development; a trade-dependent, highly industrialized system would have a rank-size distribution topped by a primate city (fig. 2.3E).

These models are still very general. There are many exceptions, such as the primacy of Paris, Rome's lack of primacy, or Auckland's increasing dominance over industrializing New Zealand. Nor do they tell us much about the spatial structures within the systems, unless they have central place patterns as in Vance's model. Many types of town in fact cluster together, especially where they are related to resource locations, as in Appalachia or South Wales; many small towns in the export specialization or technical virtuosity stages of Thompson's model occupy the zone around a major urban centre, or are strung out along a main corridor of development joining two major centres. Nevertheless, these various depictions do give some indication of the general forms which urban systems take, and the processes by which they attain them.

The system in operation

So far the emphasis has been on the components of the system, with little reference to the interactions among them, and how these are organized. Given the general model of system operation outlined in this book, however, one would anticipate that: (1) since a major raison d'être for towns is supposedly the external economies of scale, as much interaction as possible should be intra-urban and (2) because of the role of distance as a cost factor (in many senses – time, opportunity, and information as well as finance), external interaction should be over the shortest distances possible.

THE MOVEMENT OF GOODS

Relatively little can be said about the first anticipated flow pattern for the movement of goods. Karaska's (1966) analysis of the importance of intra-urban flows for Philadelphia suggests that the efficiency hypothesis is incorrect, however. He concluded (p. 368) that 'The largest inputs to Philadelphia industry are procured from non-local sources, and those largest inputs which are purchased locally are small in size.' But Philadelphia is very accessible to suppliers within the United States and with many of its products the inputs form a small proportion of the total price, for which transport costs are often minimal. Industrial agglomeration, then, need not be for material linkages, though this depends on the nature of the inputs. The city, however, is one of the major nodes of a multi-nuclear region, so one could still expect most flows to be very local. In this case, the expectations may be altered to read that: (1) most movement should be between city and hinterland and (2) outside that zone, distance should be strongly related (negatively) to flow volume, holding constant variation in supply and demand.

Flow patterns demonstrating the above general regularities have been shown for the Bristol region in England (Britton 1967). Aggregate flow data have been fitted to the 'gravity' equation, derived from Newtonian physics, whose general form is

$$I_{ij} = kP_iP_j/D_{ij}^q$$

though it is usually fitted as

$$I_{ij} = aP_i^{k_1}P_j^{k_2}/D_{ij}^q$$

where P_i = a measure of propensity to interact at the origin
   P_j = a measure of propensity to interact at the destination
   D_{ij} = distance between i and j

a, k, and q are empirically derived constants

and I_{ij} = the interaction between i and j

This equation is generally transformed to *

$$\log I_{ij} = a + k_1 \log P_i + k_2 \log P_j - q \log D_{ij}$$

though in a study where place i is always the same, that term is removed.

The gravity equation suggests that the volume of trade between two places is a function of the product of their propensities to interact (usually measured as their populations) divided by the distance between them: each term is weighted when the equation is fitted, since the equation is only a descriptive device and not an explanatory model of movement. The logarithmic* form of the equation suggests that (a) as the propensity to interact increases, the rate of increase of trade also increases, and (b) as distance increases, the rate of decrease in trade itself declines. For rail flows from Bristol to 128 zones of England and Wales the equation

$$\log I_{ij} = 2 \cdot 68 + 0 \cdot 62 \log P_j - 1 \cdot 48 \log D_{ij}$$

accounted for 42% of the variation in flows, whereas for flows by road to 107 zones the fitted equation

$$\log \text{Iij} = 6 \cdot 25 + 0 \cdot 80 \log \text{Pj} - 3 \cdot 13 \log \text{Dij}$$

accounted for 68% of the variation. The much larger regression* coefficient for Dij in the second equation indicates that the volume of road traffic declined much more rapidly with distance from Bristol than did the volume of rail traffic (see page 6). In general, the shorter the distance, the greater the cost advantage for road over rail in moving goods. This is reflected in the Bristol case by the types of commodity carried by the two modes. Road was the main choice for the movement

TABLE 2.2 *Type of commodity, location of production and distance moved: United States, 1960*

Product	RELATIVE ACCESSIBILITY OF PRODUCTION SITE[a]					
	High		Intermediate		Low	
1. Raw material-oriented Aluminium	Ohio	1. 65 2. 13 3. 22	Louisiana	1. 4 2. 94 3. 2	Washington	1. 37 2. 8 3. 55
2. Market-oriented Agricultural implements	Illinois	1. 77 2. 22 3. 1	Tennessee	1. 76 2. 19 3. 5	California	1. 20 2. 42 3. 38
3. High value added Household electrical equipment	New York	1. 50 2. 27 3. 23	Missouri	1. 95 2. 5 3. 0	California	1. 73 2. 17 3. 20

[a] The three rows in each cell indicate the percentage of all outflows moving: 1, less than 800 miles; 2, 800–1600 miles; 3, more than 1600 miles.
Source: from Pred (1964a, figs. 4, 5, 6)

of foods, textiles, clothing, footwear and engineering products, most of which are destined for the local consumer market. Rail, on the other hand, was mainly used in shipping tobacco, chemicals and paper, which mostly went to distant markets for further processing. These two different flow categories in fact reflect Bristol's two main roles in the British space economy. As a regional metropolis, it fabricates goods for the local market, based on local inputs plus others, like textiles from the North of England, in which certain other British regions specialize. Secondly, as a centre in the stage of technical and professional virtuosity, it exchanges goods—mainly by rail—with other regions. Many of Bristol's specialities in this category are based on imported materials.

Rather than focus on the place, one can also study the flows of types of goods, and Pred (1964a) has produced a classification of such trading patterns. Two criteria were used. First, the area of origin was categorized according to its population potential, as an index of its general accessibility to the national market. Secondly, the commodities were classified into: (1) those which are raw material or power oriented and produced at fixed locations, (2) those which are sensitive to transport costs and so located near to their markets, and (3) those high value added industries which cluster for agglomeration economies. A nine-cell typology was produced, and representative commodities and locations chosen (table 2.2, in which flows from each origin are grouped by percentages into those less than 800 miles, those between 800 and 1600, and those exceeding 1600).

Commodities of the first type are represented by aluminium. They are local regional specializations distributed to the national market, so the less accessible the state, the longer the average haul of its aluminium products. Agricultural implements represent the market-oriented commodities, whose movements are predominantly over short distances; California is an exception because of its specialization in certain types of implements such as cotton-picking machines. Finally, with the high value added commodity, such as household electrical equipment, the two relatively inaccessible areas, Missouri and California, produce mainly for their local markets, whereas New York, undoubtedly through scale economies and specialization, retails its products much more widely, indicative of its metropolitan function.

The total pattern of flows between the components of a system, as demonstrated in a number of studies, has been summarized by Berry (1966, 188) as

> a set of metropolitan regions within which exchanges of each area are dominantly to and from the metropolitan centers, perhaps via smaller nodes in the urban hierarchy. Each region also has certain specialties that it provides for the nation as a whole—either based upon major resource complexes, or industry in the metropolis and its satellites. Flows of these specialties between regions hold them together in a national economy, although the preponderance of them are routed between the metropolitan centers.

Clearly this views cities as the major articulators of the space economy, which apparently holds true for countries at all levels of development. In Nigeria, Hay and Smith (1970) have shown that over 80% of total inter-regional trade is handled by only 23 towns, which may mean the produce changing hands several times. And Berry's (1966) own detailed study of the Indian space economy demonstrates a similar pattern to that described in the above quotation. The country was divided into a number of separate regions according to their economic functions, notable among which were the metropolitan centres of Bombay, Calcutta, Delhi and Madras. The inter-regional flows of 63 commodities among these 36 areas were analysed to indicate the following underlying regularities.

(1) Flows focused on the four metropolitan centres, which included:
   (a) those assembled from the local region for local consumption, regional redistribution, or export;
   (b) those assembled by coastal shipping for local consumption or regional redistribution;
   (c) those assembled from national sources for local and regional consumption;
   (d) those imported from abroad for local or regional destinations;
   (e) those produced in the metropolis, for local, regional, or national markets.

(2) Flows to and from non-metropolitan areas, including:
   (a) those exchanged with the local metropolis;
   (b) those exchanged with neighbouring areas;
   (c) those involving nationwide flows of specialized products.

TRADE AND PRICE

A probable outcome of regional specialization and the consequent inter-regional trade is variation in commodity prices according to the distance between points of supply and demand. (Although in many countries uniform prices are often imposed by governments or by producers themselves.) Egg prices in Sweden are an example of this (Tegsjo and Oberg 1966). The egg supply and egg demand potential for each county was computed. Supply potential was calculated with the formula

$$SS_i = (S_i/d_{ii}) + \Sigma(S_j/d_{ij})$$

where S_i = egg production in county i
        S_j = egg production in county j
        d_{ii} = a measure of the average distance in county i
        d_{ij} = the distance between counties i and j
and SS_i = the supply potential

The higher a county's supply potential, the closer it was to large supplies and therefore the lower the egg price should have been. Demand potential was measured similarly as

$$DD_i = (D_i/d_{ii}) + \Sigma(D_j/d_{ij})$$

where D = demand for eggs, otherwise the notation is as before.

The higher the demand potential in a county, the greater the competition for eggs, and the higher the price should have been. These two measures were then used as independent variables in a multiple regression* with the price of eggs to the consumer as the dependent. The resulting equation

$$\text{Price} = 574 \cdot 4 - 0 \cdot 19 \text{SSi} + 0 \cdot 14 \text{DDi}$$

accounted for 50% of the variation in prices, with each variable operating as expected. Further analysis indicated that prices were more sensitive to variations in demand potential than in supply potential.

If the prices of all commodities vary according to their supply and demand potential, the result could be considerable inter-place variation in an inhabitant's total food bill. Gould and Sparks (1969) have reported a speculative inquiry into this in south-central Guatemala. They computed a minimum diet according to calorie, protein and vitamin requirements and then estimated its cost at various points. The average cost was 10·2 cents per day, but this allowed for minimum nutritional value only. Perhaps more interesting was the range of prices from 8·1 to 13·2 cents. Such variation remained for a variety of other diets. In all cases, the largest centre of Guatemala City stood out as a high cost location. Demand potential was high there, but the need to bring supplies long distances meant that supply potential was relatively low.

THE SPACE ECONOMY AS A CONSUMPTION SYSTEM

So far this discussion has focused on the flow of goods. But, as central place theory indicates, the system also operates with the customer going to the distribution point (unless it comes to him as a travelling shop or a mail-order catalogue). Central place theory assumes that people obtain their goods and services from the nearest alternative, thereby minimizing transport costs. Many studies have tested the validity of this assumption.

One example of these investigations is Barnum's (1966) research on the shopping habits of some 1500 inhabitants of the area around Heilbronn, a town between Stuttgart and Heidelberg in Baden-Wurtemberg. The higher the order of the good, and on average, therefore, the less frequently it was purchased, the more likely a person was to purchase it outside his home settlement—which presumably had no establishment retailing it. The higher the order of the good, also, the more likely people were to buy it at other than the nearest retail outlet, though the majority of people shopped at the closest centre. Similar findings are reported from a wide variety of other places, though each may have its own characteristics. In Australia, for example, it seems that the further small town residents live from a larger regional centre, the more likely they are to by-pass the latter on their major shopping trips, preferring to travel further to the wider choice available in the State capital (Johnston and Rimmer 1967).

An alternative examination of consumer behaviour patterns classified trip types according to the comparison between actual distances travelled and the shortest distances necessary to purchase the relevant good or service (Golledge, Rushton and Clark 1966). Among the dispersed inhabitants of rural Iowa, some goods were characterized as *spatially flexible*, because the consumers were prepared to shop around for them and travel much longer distances than absolutely necessary. Purchases of new cars and of women's clothing exemplify this. Both are commodities which vary considerably in type and quality and for which certain lines are only available from certain outlets. A farmer can probably get the latest Ford model from the nearest dealer, for example, but may have to go much further for a Mercedes-Benz. Other behaviours were categorized as *spatially inflexible*, since residents tended to use the nearest alternative. More standardized goods and services were typical of this group, such as church-going, TV repairs, and the purchase of groceries or fuel. Together these two types, plus those intermediate to them, suggest flow patterns similar to those outlined earlier for inter-regional commodity trade: local movement for general, standardized and widely available products and a wider movement for the more esoteric goods and services.

Other deviations from the pattern of consumer behaviour predicted by the central place model may result from personal differences in attitudes to and information about shopping alternatives. In Canada, for example, the members of the Old Order Mennonite religion are much more spatially inflexible than their 'modern Canadian' neighbours (Murdie 1965). In other cases some may travel further than necessary because they are not aware of the alternative, or because of their attitude to shopping. To some people shopping is a chore, to be completed rapidly, while to others it is a joy on which much time and effort may be expended.

THE SYSTEM AS AN INNOVATION CHANNEL

The role of the spatial system as a channel for the diffusion of innovations has been suggested several times in the book already. Two spatial processes of diffusion are usually hypothesized. The first suggests a distance-decay pattern,* in which the innovation spreads outwards across space from its origin, perhaps uniformly in all directions but more probably warped by the form of the transport and communications networks. Such a process clearly follows on from the distance biases in information flows, the product of the time and cost involved in communications, which were discussed in chapter 1 (p. 6).

The second spatial diffusion process suggests that the path of an innovation follows the main links of an urban system (see above, p. 27). In its ideal form this is hierarchical, and the main contacts to and from any one town are with its nearest neighbours, both its spatial nearest neighbours and those in adjacent orders of the hierarchical system. Diffusion of innovations may flow through the hierarchy, therefore. The largest centres in a system are often the main innovation centres. Inventers tend to cluster in such places (Feller 1973), largely because large cities are the chosen locations of large firms, those which invest most in research and development. From its origin and initial adoption in the largest centre, the innovation will then diffuse down into the smaller towns. This is often a financially induced process, because a place must offer a sufficiently large market for the innovation to be successfully adopted there. (This argument applies not only to technological innovations, gasworks for example, but also a wide range of social and cultural changes, such as the introduction of discotheques.)

Together, these two processes suggest an orderly procedure by which innovations are diffused through an urban system. From their origin, usually in one of the larger towns, they would be adopted in places which are (a) spatially close to the origin, (b) near to the origin in size. Such places are the most likely to obtain sufficient information about, and then to adopt, the innovation. These places would then act as diffusion centres too, so that eventually the innovation reaches the smallest and most isolated components of the spatial system.

One of the clearest examples of the general spread mechanisms in operation is in Pred's (1971a, 1971b) study of the diffusion of news items in newspapers through the United States in the era prior to electronic communication, when virtually all information was transferred from person to person by written material or word of mouth. Maps of the time-lag of news for New York at three separate dates (fig. 2.5) show a clear distance bias—which contracted over time as modes of transport accelerated. In general the lag was much greater to inland centres than to places connected to New York by the more rapid and regular coastal shipping, though some inland contacts to major centres (New York to Albany, for example, or Philadelphia to Pittsburgh) were relatively quick. Within the system the most frequent and rapid contacts were between the major centres of Baltimore, Boston, New York and Philadelphia, and because of this the flow of intermetropolitan news dominated the slower rate of diffusion between metropolis and hinterland. A similar communications network exists in the present American banking system (Duncan and Lieberson 1970). New York dominates the country; each region is dominated by a regional metropolis; the regional metropoli and New York are linked together in a countrywide communication system.

Many other innovations follow similar paths through the spatial system. The first American television system, for example, opened in the second largest city in 1940. New York was the location of the next,

D

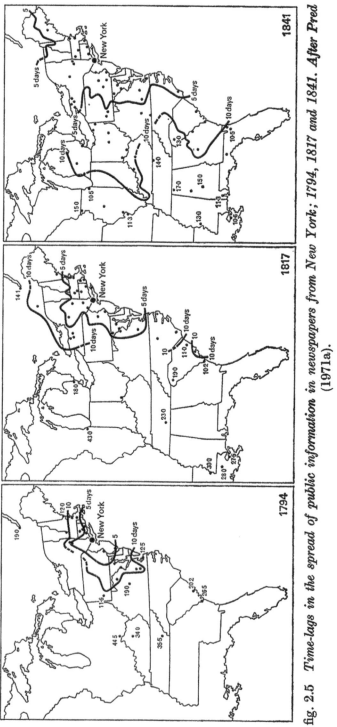

fig. 2.5   *Time-lags in the spread of public information in newspapers from New York; 1794, 1817 and 1841. After Pred (1971a).*

in the following year, and from then on there was a general trend of the smaller the place, the later the arrival of a local TV station (Berry 1971b). Furthermore, the first small places to get a station were close to a larger and earlier adopter. (It is interesting to note a reverse pattern, from suburb to city, for colour TV: Berry 1971b.) This process of hierarchical diffusion has been detailed in a study of the spread of planned regional shopping centres through the USA (Cohen 1972), relating the date, magnitude, and intensity of adoption of this institutional innovation in retailing to a range of variables such as the size of the local market and the amount spent on general merchandise. In general, the less that was spent per head on retail goods, the lower was the intensity of adoption (measured as square feet per capita), and the market variables accounted for most of the variation in this dependent variable. For the magnitude of adoption, measured by the size of the shopping centres, urban size was a basic criterion, though a strong CBD might act as a delaying factor; the later the adoption, however, the weaker were the observed relationships. Finally, the independent variables representing potential markets were least successful in accounting for variations in the date of adoption of centres; basically, however, the larger metropoli adopted first.

These examples, a selection from a large literature on the diffusion of innovations through spatial systems (Brown 1968), clearly validate the general model of two consecutive processes, the spatial spread and the hierarchical effect. But there are many exceptions to those 'rules', particularly that referring to hierarchical patterns. Innovations requiring little capital investment may arise virtually anywhere in a system, wherever the 'inventer' might live. That place may retain its primacy for that 'invention', as was the case with the industrial revolution on Britain's coalfields. Often, however, it may be necessary for an inventer to move to a large place once he wishes to market his product widely. The first jet-boats in New Zealand were built on a High Country sheep station; the inventer now manufactures them in the South Island's largest city. Large centres may repulse innovations, perhaps because of their existing fixed capital in the commodity which the innovation might replace, in which case the entrepreneur might select a smaller centre. With the increasing role of large corporations and of state enterprises in technological developments, initial adoption is often experimental and may be in a remote part of the system. Britain's first motorways, for example, were in Herefordshire and in North Lancashire. And the same is true for other innovations; many new stage-shows are performed first 'off Broadway'. But in such cases, successful innovation in the provinces is usually followed by introduction to the largest centre from which the hierarchical and spatial spread processes emanate. The first 'major' motorway in Britain linked

London with Birmingham; Broadway in New York and the West End in London are the main areas for initial theatrical success. Finally, the processes of spatial diffusion identified here may be distorted because the innovation is aimed at a certain section of the population only, who may not live in the system's metropolitan centres. The spread of the Co-operative movement from its origin is Rochdale, Lancashire, is an example of this.

The spatial system may channel a wide variety of changes, not only those connected with the diffusion of innovations. Business cycles, as reflected by unemployment rates, may display spatial dimensions, with certain places 'leading' while others 'lag' behind, and with groups of adjacent places experiencing similar patterns of temporal change. Often these spatial groupings result from the linking of adjacent places into industrial complexes. The New England textile towns, for example, have all recently experienced increasing unemployment, in addition to the national trend in the United States, which reached a peak of unemployment in 1962 (Casetti, King and Jeffrey 1971). In the American Midwest, during the period 1960–4, there was a general trend in monthly unemployment rates plus a specific trend relating to each of five groups of places organized around the metropolitan centres of Cleveland, Pittsburgh, Detroit, Indianapolis and Chicago (Jeffrey, Casetti and King 1969).

The changing spatial system

The changes discussed in the previous section are short-term only, and usually produce only minor adjustments in the spatial system, in the size and functions of its component parts and the interactions among these. But these alterations occur within a matrix of long-term change, brought about by improvements in transport and communications which are major spatial elements in the modernization process. These improvements have already been termed time–space convergence (see p. 9).

Three separate spatial trends have been recognized as usual concomitants of, as well as frequent preludes to, further time–space convergence (Janelle 1969). The first, probably the major, is *centralization*, by which activities concentrate into a few centres, better to reap advantages of scale economies. If time–space convergence occurred at a uniform rate over a space economy, any place might be chosen for the centralization of activities. But two related considerations oppose this. First, at any one point in time some places are more accessible to a total system than are others. They build on this initial advantage in their larger size and existing ability to benefit from scale economies. Second, the rate of time–space convergence is usually far from uniform.

Some places are preferred to others in the development and/or upgrading of transport and communications media. Larger places are usually the first to be connected, since the greatest returns from new routes come when they join the greatest traffic generators (Lachène 1966): hence, initial advantage is cumulative.

Among a sample of 13 SMSAs in the US northern Midwest, the average convergence, minutes saved per route mile travelled, from Detroit to the other twelve was 0·513. For Chicago it was 0·486; for the next largest, Grand Rapids, 0·422; and for Port Huron, the smallest, the average convergence was only 0·338 (Janelle 1969). Detroit's relative accessibility increased considerably, therefore, since it came 'closer' to the total system than did the others. This trend was matched by a growing centralization there of the wholesaling industry for the area. Similar trends are apparent in many industries. Brewing in New Zealand, for example, was widely distributed in the late nineteenth and early twentieth centuries, with each local plant having a monopoly over a spatially circumscribed area. Transport costs protected such monopolies, but as the costs were gradually reduced so the larger breweries of the big cities were able to take-over their smaller competitors. This established a new spatial equilibrium, comprising, in the main, a few major breweries in the biggest urban areas, each with an extensive spatial monopoly (Golledge 1963).

Individual residents of a system have also become more mobile, tending to travel further, for example, on shopping trips. Comparative analysis of the trips undertaken by rural Iowans in 1934 and 1960, found that at the latter date people shopping for groceries were much more likely to substitute a larger centre, further from home, for the small, local centre than was the case three decades earlier (Rushton 1969). The longer the distance to the nearest small centre, in fact, the more people were prepared to travel even further (see also the earlier discussion of Johnston and Rimmer 1967).

These changes in consumer spatial behaviour have had obvious consequences for the central place system. In Saskatchewan, this has been represented by a decline in the number of small centres, except the smallest whose numbers have increased as many formerly more important centres have 'regressed' through the hierarchy (Hodge 1966). Such local service centres are no longer viable. Many of their hinterland residents move away, as part of the general rural depopulation. The reduced market means that fewer services can be provided in the centres and greater mobility leads the remaining population to shop in larger, more distant towns. These trends, which are particularly marked close to the main centres, result in some of the larger towns moving 'up' the urban hierarchy (table 2.3). The result is a denser network of major centres (as shown by the declining average distance between them), a lesser density of middle-order centres, and perhaps a slightly greater density of the smallest, most basic settlements.

Other changes in the system accentuate these trends. Economic progress is characterized by growing proportions of the labour force in the more footloose tertiary and quarternary industries. In fact, these are already very centralized. With the greater concentration of industry into fewer firms, this trend is being magnified. Improvements in technology mean that while some industries become more footloose, usually to concentrate in some major centre, others are more tied to certain type locations (Beaver 1961). Ever more complex processing and fabricating procedures lead to increasing inter-plant linkages and interdependencies (Rose 1969), all of which encourage economic and population centralization.

Deconcentration, the second of the processes, is the product of two sets of forces. First, many large and rapidly growing places suffer from

TABLE 2.3   *The changing central place hierarchy of Saskatchewan, 1941–61*

Order of centre		Percent change Number of centres	Percent change Mean distance between centres
Largest	1	0	0
	2	+80	−44
	3	+12	− 2
	4	+50	−14
	5	−40	+22
	6	−48	+31
Smallest	7	+12	+ 5

Source: Hodge (1966)

congestion which may, at least at the local level, produce time-space divergence rather than convergence. Other deleterious trends, such as various forms of pollution or land value inflation, may accompany this congestion. Secondly, the greater accessibility puts many places as much as 50–100 miles distant within easy reach of a major centre, so that they become attractive locations for many functions. In California, for example, the rate of increase in employment among thirteen cities between 1951 and 1967 decreased with increasing distance from Los Angeles (Casetti, King and Odland 1971).

Suburban and ex-urban sprawl are caused by these centrifugal forces, acting at a more limited spatial scale within the pattern of operation of the forces of centralization. Such sprawl typifies many major metropoli, around which people and jobs are being diffused. Many new jobs have been created in recent decades in London's peripheral zone. The percentage rate of manufacturing growth in this area among the

New Towns was negatively related to distance from a trunk road and to population potential, suggesting the importance of accessibility to the location decision-maker, and was positively related to the percentages of skilled and unemployed workers there, indicating the attractions of certain types of labour market. Among the other districts, percentage growth was greatest where manufacturing was relatively strong, which is in line with hypotheses concerning the importance of external scale factors. If an absolute measure of change over the period is used as the dependent variable, then among the New Towns the 'growth creates more growth' syndrome stands out, but among the other areas labour availability, a favourable existing industrial 'mix',* and access to London are the apparent determinants of expansion (Keeble and Hauser 1971, 1972). Overall, deconcentration is channelled to certain places, with accessibility being an important influence on the routing system.

The final trend associated with time-space convergence is *specialization*. Although accessibility improvements generally favour centralization into the larger places, they do benefit all settlements to some extent. Thus an area with a comparative advantage for certain agricultural production, or a town with a well-run factory, can widen its selling area and thereby perhaps come to enjoy scale economies. This may result in a new regional metropolis—as occurred on the west coasts of the USA and Canada. More likely it will allow a smaller centre, perhaps even a very small one, to enter the fifth stage of Thompson's model (Professional and Technical Virtuosity) without first passing through the third and fourth. In effect, specialization is a more limited form of centralization, the difference being in the number of functions which become concentrated within one place.

This process of regional specialization combating overall centralization is occurring in New Zealand at present. Industrial growth, especially that concerned with the national import replacement process, is very concentrated in Auckland, the largest city. But other centres are developing certain particular roles in the system, often based on local resources, as with forestry in the North Island and an aluminium refinery based on cheap power in the South, with fruit and vegetable processing on the east coasts of both islands, and with tobacco manufacture in the Nelson region. Others base their speciality on industry, as with car assembly at Hutt and the oil refinery at Whangarei (Johnston 1971b). Such growth enables further local import replacement industrialization, of course, especially in such labour-oriented activities as clothing manufacture. It is probable, though unlikely, that it may generate enough development to halt the centralization trend. In most situations, however, the latter seems permanent, despite many views to the contrary.

CHANGE IN SUMMARY

The morphology of a space economy is a compromise between two opposing sets of forces; the scale economies—internal and external—of concentration, and the locational costs of isolation from either supply or demand centres. In the early stages of modernization, the latter force often dominates in moulding the spatial system. Population and production are widely scattered; interaction is slow, costly, and slight. With technological advance, especially in transport and communications, the balance tips more and more towards the first set of forces, as detailed above, though the concentration that it produces is often fairly diffuse and there is sufficient momentum in most systems for at least some places to contest the growing centralization.

Fortunes of urban places, illustrated by their population growth rates, outline the general trends within such a modernizing system. In the United States, for example, it was hypothesized that the smaller the place in 1940, the more likely it was to decline over the succeeding two decades (Northam 1969). Similarly, the further a place was from an Urbanized Area (a census definition: see Murphy 1966), the greater the probability of decline. Data indicated that the latter hypothesis was certainly valid, the former generally so; jointly the two variables were 86–96% accurate in forecasting urban fortunes in three study areas. In Australia too, where a study of all towns with less than 25,000 inhabitants in 1961 showed that in the following five years:

(1) there was a lower probability of decline among those with 5000 or more residents than among all towns;

(2) there was a lower probability of decline among towns close to metropolitan areas than among all towns;

(3) there was a lower probability of decline for towns in areas where the rural population was increasing.

Exceptions to this set of rules highlighted those small towns where specialization promoted growth; notable among these were the tourist and resource centres (Johnston 1968b).

A final illustration of the three trends of centralization, deconcentration, and specialization is given in a model of port development applied to both Australia and New Zealand (Rimmer 1967a, 1967b). Initially there were many small, independent ports arrayed along the coast, but some were soon able to develop hegemony over others, creating a hierarchical system (fig. 2.6), through their control of inland and coastal transport. By decree, the location of government facilities such as customs posts, chance, or superior facilities, centralization into the favoured ports continued. Some ports (P 4 in fig. 2.6, for example) may develop special facilities to counter such a trend—

perhaps an oil terminus or loading gear for frozen carcasses. And finally, alongside the centralization and specialization comes deconcentration: congestion in the major port leads to certain functions, especially those concerned with bulk cargoes, being translated to adjacent subsidiary centres.

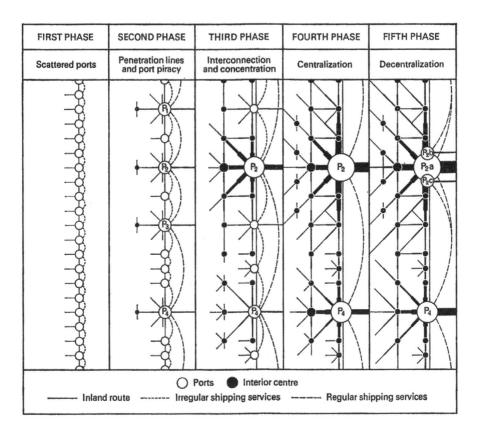

fig. 2.6 *Development of a hypothetical port system. Source: Rimmer* (1967b).

Migration and the space economy

Modernization is both cause and effect of urbanization, or population concentration, and, as outlined above, with continuing development this usually involves growing centralization. Gibbs (1963) has suggested five separate stages to this:

(1) Rural growth rates exceed those of urban places.
(2) Urban growth rates exceed rural.
(3) Rural population experiences absolute decline.

(4) Small towns decline, relative to larger centres if not absolutely.

(5) Large towns spread (deconcentration).

These clearly parallel the transition from an economy dominated by primary occupations to one whose tertiary sector is the largest. Such differential growth rates may result from variations in rates of natural increase. In the nineteenth century urbanization of Europe and America, however, migration was dominantly the major factor in urban growth. The more recent urbanization of the developing countries has shown a larger relative proportion of growth by natural increase (Davis 1965): nevertheless, the volume of urban-directed migration streams in these countries is very great.

Given that migration is a major cause of variations in growth rates, one can deduce from Gibbs' model a process of movement from rural areas to small towns, and then from those places to the larger centres. The spatial order in such a pattern was suggested in 1885 by Ravenstein who, from a study of birthplace data, proposed a series of 'Laws of Migration' including the salient elements that:

(1) Most people move only short distances.

(2) Growth proceeds by towns attracting migrants from their immediate hinterlands, which in turn make up their losses by attracting migrants from more remote areas. From the towns, migrants proceed to larger centres, so that the basic movement pattern is upwards through the urban hierarchy. (Note that this is the reverse of the flow of information in a system, detailed above—p. 38).

(3) Long-distance migrants generally proceed to a large place. From this general pattern there are, of course, many deviations, including counter-currents whose strength relative to that of the main current may vary considerably (Lee 1966).

Ravenstein's 'laws' suggest the attractions of size and the frictions of distance for migrants, as for the movement of goods, and the gravity model formula has been fitted successfully to this flow pattern many times. Assuming that all people are equally likely to move, then the larger a place the greater the number of potential migrants, in both directions. Distance is a friction effect, representing the influence of spatial variations in travel time, cost, information and opportunities. The further away a place is, all other things being equal, the less people are likely to know about it and so the less likely they are to move there. The distance-decay* in information levels is likely to be very rapid over the first few miles from the origin, though less so as transport technology improves. Several recent studies of 'mental maps', of people's knowledge of and attitudes to their spatial environments, have demonstrated such a distance–decay pattern. Around the home town, there is a very clear preference for local areas, but this declines rapidly into a wider area in which distance is but very poorly related to preferences for places as living areas (Gould and White 1969; Johnston 1970). Hence the distance-decay in migrations is exponential in form, with a rapid decrease in volume of movement over the first miles, but with this rate of decrease falling. Finally information levels probably cause the relationship between length of move and size of destination. Of two towns which are both equidistant and a considerable distance from a potential migrant's origin, more is likely to be known about the larger of the two, as activities there will be more commonly featured in the mass media. Assuming that people will tend to move to places they know most about, therefore, the larger place is the most probable destination. It will probably also offer the wider range of opportunities.

Many researchers have fitted the gravity model equation to the total pattern of migration in a system. Schwind (1971), for example, was able to account for nearly 60% of the variation in total flows between major urban regions in the United States with the equation

$$\log \text{Mij} = -2{\cdot}3 + 0{\cdot}64 \log \text{Pi} + 0{\cdot}71 \log \text{Pj} - 1{\cdot}11 \log \text{Dij}$$

where Mij represents total migration between i and j and the other notation is as above (p. 33). Nevertheless, critics are often dissatisfied since the gravity model is merely an analogue derived from physics and does not 'explain' why size and distance should be exponentially related to total migration volume. A more realistic model was formulated by Stouffer (1940, 1960) who replaced distance by 'intervening opportunities'. The greater the number of opportunities between A and B, and the smaller the number of competitors for those, the smaller should be the volume of movement from A to B. Stouffer was unable to measure intervening opportunities and competing migrants properly, however, while later work (Haynes *et al.* 1973) has sometimes been unable to discriminate between the two models in terms of their predictive power.

Another drawback of the gravity model is its often poor predictive performance when gross flows between places are studied. For this reason, the regression equation has been successfully expanded to include such variables as differences in population growth rates, wage rates and unemployment levels. One such analysis suggested that migration flows in the United States comprise three component patterns (Schwind 1971):

(1) Long-distance shifts from all parts of the country towards the attractive environments and booming economies of California and Florida.

(2) Medium-distance shifts from rural and small-town hinterlands to regional metropoli.

(3) Deconcentration, from city to suburb.

These three elements closely parallel those in the movement of goods (p. 36).

Such findings suggest that Ravenstein's 'laws' are still of general relevance, especially when the distance and size variables are incorporated with a model of career cycles. Study of migration in a part of New Zealand produced the following threefold classification of migration types (Keown 1971), whose spatial correlates are similar to those documented by Schwind.

(1) Local migrants within the rural economy who change their job, but not their occupation, in a search for better opportunities. The major influence on their movement is the distance variable.

(2) Occupationally mobile people, who move into the better social and economic opportunities of the urban environment. Both urban size and distance influence their movement; most go to the nearest large centre.

(3) Career transients, most of whom are members of organizations, or who have occupations, for which job information is nationwide. They move, or are moved, throughout the system, with little reference either to distance or urban size; a small place at the opposite end of the country may offer the only chance of promotion. Civil servants, 'organization men', priests, and a wide range of other occupations may now conform to this pattern rather than to any of the Ravenstein laws. Most evidence suggests, however, that the metropolis-hinterland system is the main component of migrations still (Brown, Golledge and Odland 1970).

MIGRATION AND INCREASING MOBILITY

As communications improve and horizons widen, people should migrate further. This has been recently demonstrated by Tarver (1971) who estimated the influence of distance on movement with an 'indifference' model, which assumes no constraints on movement. If the total number of migrants to state A equalled 9% of movements from all other states, then it should have received 9% of the outmigrants from each origin. On the basis of this assumption it is possible to compute the average distance migrants should travel. For all inter-state migrations in the United States between 1935 and 1940, the average distance moved was predicted as 888 miles. The average actual distance moved was only 68% of this; migrations were in fact biased towards the nearer opportunities.

From the late 1930s to the late 1950s, average migration distances have increased by some 25%, from 606 miles during 1935–40, through 689 in 1949–50, to 756 in 1955–60. In the first period, over half of all inter-state moves were less than 400 miles in length; twenty years later the relevant mileage was 600 (Tarver and McLeod 1970). The ratio between actual distances and those predicted by the 'indifference' model has also increased, from 68% in 1935–40, through 70% a decade later, to 79% in 1955–60. Clearly, American migrants are becoming more indifferent to the costs of overcoming distance, perhaps because of the growing proportion in the national system who are career transients.

Structure, process and life style

The role of the city as a centre of social as well as economic change, and as the source of information about such change, has been pointed out in the first chapter. Various levels of economic development, it was suggested, are associated with many aspects of life style. This relationship between the structural and behavioural categories of urbanization has been related by Wirth (1938) to the demographic category, to the increasing population concentration which accompanies modernization. Urban population size, density and heterogeneity are presumed to be strong influences on a wide range of inter-personal relationships, hence hierarchical variations in life styles are to be expected (or, if a hierarchy is not present, a rural-urban continuum).*

While such patterns may be seen in one temporal cross-section, life styles are often rapidly changing, with most developments emanating from the largest cities, the main clusters of innovative minds, for whom life in group anonymity is necessary for social experimentation. Hence, over time, life style changes may be expected to diffuse down the system from the largest to the smallest settlement in a reverse flow from the upward movement of migrants detailed in the previous section. Many social innovations do not require 'markets' in the sense of scale economies, so with the widespread instant communications available in many societies, such hierarchical diffusion may not exist. Where, however, diffusion is at least partly dependent on inter-personal contacts, a hierarchical pattern following the information dissemination patterns already outlined seems likely. In addition, of course, distance can be hypothesized as a further influence, since it too is related to the volumes of goods, people and information flowing between places.

The concept of metropolitan regional dominance over the dissemination of life style changes has been studied many times, often using human fertility as the dependent variable, presumably as a surrogate for a major life style change. Kruegel (1971), for example, has investigated changing fertility levels (the number of young children per 1000 fertile females) among Kentucky counties over the quarter-century preceding 1965. Each county was classified according to:

(1) whether or not it contained a metropolitan area;
(2) if not, its distance from a metropolitan area, in three zones;
(3) its relative location to metropolitan areas as one of:
    (i) subdominant – containing a large, non-metropolitan city;
    (ii) intermetropolitan – on a major highway;
    (iii) local – all other, relatively isolated, counties.

In total, six type areas were defined and for each, fertility rates, standardized for migration, age, sex and colour variations, were computed for each county. Using these, it was hypothesized that over the 25 years rates would converge on those for the metropolitan areas, giving an urbanization of the countryside. This was confirmed. In 1940 rates ranged from 71 for the most urbanized types, through 83 and 97 to 100 for the most isolated; by 1964 rates ranged only from 94 to 99.

Similar findings have been reported for a wider range of variables over a larger area, in which counties were further classified as to whether they were in a southern state (Tarver 1969). For the educational level of females, size and location were both relevant in the south, but only distance had even a weak influence elsewhere. Similar relationships held for female work force participation rates. For fertility, however, both variables were associated with the dependent variable in the north, but only size was relevant in the south. The two general diffusion patterns, down the hierarchy and across distance, seemed generally valid then, both for these and a wide range of other variables, such as the timing of the decision to integrate schools in the southern USA, which was related to county size, and proximity to the Mason–Dixon line (Florin 1971). But the actual process is not shown. With regard to fertility changes in Sweden, for example, Carlsson (1966) has suggested that the downturn came simultaneously in all areas; areas differed at any one time because they started at different levels. If this is so, the researcher must determine how these initial differences came about.

Not every innovation originates from the major, or even a large, city; some may disseminate from the opposite pole of the spatial system. Divorce law reform and female suffrage for presidential elections within the United States both spread from the state of Wyoming, displaying a clear distance effect (Gould 1970b). No simple reason for these can be given in terms of system operation, but for others their 'backblocks' origins are not surprising. Early twentieth century agrarian riots in Russia originated in the more depressed peasant farming areas of the southern Ukraine and of the Baltic provinces, spreading towards the capital from there (Cox and Demko 1968); indeed McColl (1969) has suggested that such isolated areas, in effect the periphery of the space economy, are the usual first battlegrounds for insurgent activity. Other innovations may originate in a variety of locations, and diffuse from these according to the two usual channel types. This is so for many disease epidemics, as with the British Asian 'flu outbreak of 1957, which had origins in the east Yorkshire ports (Hunter and Young 1971). Indeed many epidemics are introduced to a country through its ports, which are often also its main cities.

The hierarchical diffusion of some innovations, notably those involving life styles and attitudes, is often impeded by the distortions in the hierarchical system of central places imposed by the industrial system. Towns of different functional types tend to have different population structures, which may influence a variety of attitudes. For example, Florin's (1971) study of adoption dates for school integration in the southern USA indicated that in addition to the size and distance effects (the latter being represented as accessibility to 'northern' attitudes), communities with large proportions of their workforces in educational or scientific occupations also tended to be early adopters, presumably because of liberal attitudes.

These findings suggest relationships between the functional character of towns and various aspects of their social structures. Such have been demonstrated in a number of analyses using the factor analysis family of techniques,* for countries at several levels of economic development. Classic among these studies is Moser and Scott's (1962) work on British towns. In this, fifty-seven variables were reduced to four main components of covariation, which were identified as dimensions of social class, housing conditions, a contrast between new development and level of labour force participation, and a contrast between growth during the period 1931–51 and certain urban characteristics (such as proportions of old persons, illegitimate birth rates, proportions of single person households, and proportions of dwellings in bad repair). Population size was not related strongly to any of these four components.

On the basis of their scores on each of these four components, the 157 English and Welsh towns were classified into fourteen groups of similar centres (London and Huyton-with-Roby did not enter any group). Members of almost all groups have either strong functional or strong locational commonalities with the other members, suggesting that the aspects of life style measured by the 57 variables tend to be distributed among the towns according to their roles within the space economy. Groups of towns with clear functional similarities included the major seaside resorts (high on social class, low on labour force participation), the main inland spas and professional/administrative centres (also high on social class, as well as labour force participation), the main railway engineering towns (generally with poor housing conditions), the main ports, the Yorkshire and Lancashire textile towns and the metal manufacturing centres. Other groups separated out the various types of suburban administrative district, with some of them clustered in certain sectors of outer London.

Such results, and similar ones from a variety of countries, India (Ahmad 1965) and other 'developing' states included, indicate that city size is but one determinant of general life styles there. Certain behaviours and attitudes are more commonly associated with certain population groups, and because the functioning of the space economy results in the concentration of many of the latter into certain places, then towns with different functions tend to have different social characters. As life styles change and are diffused through the system, this functional pattern will cause deviations from the simple hierarchy and distance based model outlined here.

Conclusions

As depicted, a national system comprises a number of places performing both complementary and competitive roles in providing goods and services for the total population. These symbiotic and competitive functions are arranged across the spatial surface in an efficient manner, in order to reduce, if not minimize, the costs of movement from place to place necessary for the distribution of goods and services to all communities. Thus, each rural area tends to specialize in that form of primary production which best meets the needs of the national, if not the international market, in a competitive situation. In addition, the immediate hinterland of most towns provides certain produce for that local market only, in a situation of spatial monopoly. Similar patterns exist in the production and distribution activities within most sectors of the economy.

This spatial order is never complete, of course. The minimization of transport costs is not a total or only goal in most societies, so decisions are often made which are 'irrational' in terms of any normative model.* Especially is this so because of the continually changing nature of most spatial systems. Many location decisions, or movement decisions, must be made in relative, if not complete, ignorance of all opportunities or of the decisions which others will make. Hence a large number of decisions may be 'conservative' in nature, following established paths rather than pioneering what may emerge as a more efficient long-term solution (Webber 1972).

Development processes, as we have seen, tend to produce a hierarchical type of spatial structure, albeit one in which the true hier-archical form is disturbed by variations in the socio-economic and physical environments. This hierarchical structure serves as a second major determinant of the spatial order in the system, along with the distance variable. Together these two influences channel many pro-cesses. The movement of goods, the flow of information, the diffusion of innovations, the long and short-term migrations of people, all tend to follow similar courses, emanating from or being oriented to the major urban places, with the secondary centres playing intermediate roles in this articulation of the system. It is in these ways, within the context of the local social, economic, political, and cultural systems and the various determinants of individual behaviour, that the pattern of places is structured into a system of national spatial organization.

3 The urban place as a system

Each urban place is a very dense concentration of people and homes, plus their attendant workplaces, centres for trade, entertainment and recreation, and the many facilities and utilities needed to service all of these. In order that the city or town may operate efficiently, some sort of spatial ordering of these functions is necessary. (The fact that numerous cities operate far from efficiently, in certain aspects – e.g. traffic congestion – indicates that the provision of spatial order has lagged behind the demand for it.) Activities must be spatially distributed so as to allow ease of access among the various components of the system, especially those which are most closely linked. Usually this results in almost complete segregation of uses one from another, when the city is divided into a fine mesh of districts, and it is a general principle of town planning practice in most countries that such segregation should be emphasized. Rarely, however, does it result in all activities of the same type being in one place: usually there are several similar clusters in different parts of the urban area.

There is, of course, a threshold size below which no spatial order is necessary. No data are available to allow any precise statement on what this minimum size is, but it is probably very small, at least in the 'modernized' societies. Settlements of only a few hundred inhabitants may have their shops, homes and places of employment located in apparently random fashion, but places need not be very much larger before a clear shopping district emerges and the various socio-economic groups come to occupy different residential areas. Such land use separation assists efficiency. It may be, however, that there is a maximum threshold size for cities, beyond which a law of diminishing returns begins to operate and marked declines in efficiency become noticeable. Again, there is no consensus on what this size may be, though it certainly will be a range rather than a specific population, since each city's peculiar characteristics will also affect its operating efficiency. Indeed, some commentators believe that there is no such threshold, that efficiency never declines with increasing size. This view involves a concept of net efficiency. Larger cities may produce increasing traffic congestion, but the costs of this do not outweigh the other benefits of greater size, such as agglomeration economies. Such cost-benefit analysis is extremely difficult to conduct, however (Neutze 1965).

In the preceding chapter, the stress was on spatial order in urban systems being the reaction of decision makers to the influence of distance on various forms of interaction. The same basic thesis characterizes this presentation, suggesting that the spatial structure of the city, and of its component parts, is a product of the same forces. The next section outlines the bases for the spatial separation of land uses, and this is followed by discussions of each major land use component.

Accessibility and spatial structure

The fundamental role of accessibility in shaping a land use pattern, via the pattern of land rents, was stressed by von Thünen in his pioneer study of agricultural location: his argument has been adopted for the intra-urban case (Alonso 1964). The analysis assumes that different sorts of firms, plus households, obtain different levels of benefits, measured in economic terms, from various accessibility levels. In its simplest form, the model can be applied to two competing land uses in a linear city (fig. 3.1A), which has one point that is closest to all the others. (This argument can be substantiated by applying the potential model introduced in chapter 2.) Given an even distribution of population through the linear city, this most accessible point will be in the centre: to reach it from elsewhere involves transport costs. Thus, if one of our two land uses was the distribution industry, the further its members were located from the central point, the greater would be their transport costs to serve the market (and so, presumably, the lower their profits). From this, the more central the location, the more valuable it is to a distributor, and the more he is prepared to pay for it, though not more than the savings he obtained in transport costs. (This model assumes no spatial pattern in any other costs.)

The second land use in the model is homes, whose breadwinners work in the distribution industry. For these too, the closer they live to the central point, where their jobs are theoretically located, the lower will be their transport costs in commuting. But these households are probably less sensitive to transport cost variations than those in the distribution trade, perhaps because they do not mind relatively short journeys, so the central land is not as valuable to them. (If it were, and they outbid the distributors for the central location, the latter would have to site elsewhere and the whole pattern of commuting costs would change. Thus households are 'followers' in the location process as described here: their valuation of sites is dependent on the valuations of others.) Figure 3.1 illustrates the values given to the various locations by the two uses, with the inner portion being bought by the

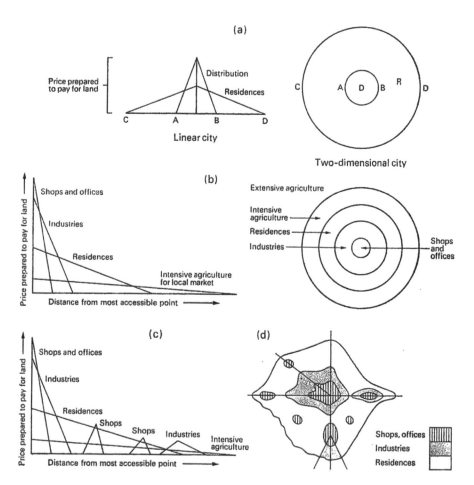

fig. 3.1 *Hypothetical intra-urban land use patterns based on rent theories.*
A. The pattern for two uses – distribution and residences – in a linear and a circular city. **B.** *The pattern for four uses.* **C.** *The pattern with local intra-urban variations in the desirability of sites, with several nodes of more intensive land uses.* **D.** *A similar pattern as C, also introducing the influence of differences in accessibility patterns related to major routeways.*

distributors. In a two rather than a one-dimensional city, this would result in a concentric pattern of land use zones.

This model can be applied to any number of land uses competing for the scarce resource of accessibility: fig. 3.1B illustrates the probable pattern when there are four. Commercial uses come into most contact with the general public. Any relatively inaccessible shop, unless it monopolizes a fairly inelastic market, usually fares badly in attracting custom. As a result, shops, and also offices, bid the highest prices for the most accessible land. (Not all shops bid as highly, however, as the ensuing discussion shows, so that there is a gradient of bids, and thus of land values, within this inner zone.) Industries only come into contact with each other, with their workers, and with their suppliers and markets, the last two often through intermediaries such as wholesalers. Their accessibility requirements are not as great as those of commercial establishments, therefore, although they exceed those of residences. As a result, industries and residences respectively occupy the second and third zones. Finally, access to the urban market is not sufficiently important for agricultural uses to outbid the others. Thus the former occupy the outermost zone: as the previous chapter showed (p. 25), this zone can be divided into sub-zones too.

The zonal pattern suggested in fig. 3.1B is characterized by a rapid decline in land values in the immediate vicinity of the central point, with the rate of decline decreasing further out. This model has been tested for Topeka, Kansas, relating land values ($ per square foot), Y, to the reciprocal of distance from the city centre, X, the reciprocal best representing the type of curve described here.* The resulting equation (Knos 1968)

$$Y = 1691 \cdot 1 + 19,975 \cdot 8X$$

accounted for nearly 64% of all the spatial variation in land values.

GENERALIZING THE MODEL

In outlining the competition for locations among land uses, the above model is extremely simple. Nevertheless, it is possible to relax its many implicit assumptions and indicate its general relevance. Competition for land is not the only process. Many firms make their location decisions after viewing the actions of others: shopping centre development, for example, is usually undertaken by a large store, and the little ones accept the logic of the choice and move there too, to benefit from the shoppers drawn to the centre by the large store. Similarly, households often follow the dictates of real estate agents. In these ways, an existing pattern becomes deeply ingrained, and there is considerable inertia, particularly where the land is owned by the user. As a result, the 'real world' is usually not as simple as the competition model proposes.

It is difficult to incorporate factors of inertia and 'locational decision-making sheep' into the model. Other of the model's assumptions can be successfully relaxed, so that it more closely approximates reality. For example, not all firms are competing for central locations. Among commercial users, some will require access to the whole city, and the area beyond it; others only need to be close to a portion of the total market. The former will in all probability have to locate in the central portion; the latter may be able to obtain a spatial monopoly over part of the city from a non-central site. Industries, too, may vary in their locational requirements, producing the sort of multi-nodal pattern in fig. 3.1C.

A concentric zonation of the city (fig. 3.1B) is only possible if movement is equally easy in all directions. Within a city, however, all interaction is channelled along particular routes – notably the street system – so that in some directions from the centre accessibility is greater than others (fig. 3.1D). The concentricity of the zones is then destroyed, with some of the major routes attracting land uses. With very complex transport patterns, the zones may be so distorted that the accessibility component in their formation is very hard to discern (Quinn 1940).

Knos expanded his equation describing land values in Topeka to take these two distorting factors into account. Three further variables were included: distance from the main thoroughfare, population potential, and the direction of urban growth. Together with distance from the city centre, the total spatial variation accounted for was 79%, as compared to 64% for the distance from the city centre variable alone. All three extra variables made a significant contribution to the equation, though that for population potential was slight.

One assumption not yet relaxed is that accessibility, and hence transport costs, is a major influence on location decisions. With time-space convergence and increasing levels of car ownership, many people are now indifferent to location relative to the city centre, particularly since the transport system at that point is often inadequate, producing a local time–space divergence. This could lead to a decline in spatial order. More likely it produces a new form of order, as Yeates' (1965) study of Chicago land values shows. He related front-foot land values during the period 1910–60 to six independent variables: distance from the Central Business District (CBD), distance from a regional shopping centre, distance from the nearest subway, distance from Lake Michigan, the percentage of the population who are non-white, and population density. The first three of these relate to the model shown in fig. 3.1D; the next two represent social evaluations of properties; the last is really a consequence of land values, since the more expensive an area is, the more it is likely to be subdivided among many residences. Over the fifty years, the six variables accounted for an ever-decreasing proportion of the spatial variation (from 77 to 18%) as land values became less sensitive to the various influences. Furthermore, the relative importance of the variables changed. In the early years those representing the model of fig. 3.1D predominated; by 1960 it was the social evaluations which were more important and the influence of accessibility had declined considerably.

Finally, the model outlined here assumes that the various land uses – commercial, industrial, residential – are homogeneous categories. They are, of course, composite terms which cover a wide range of firms in each, which may have different locational requirements and hence produce spatial patterns, or structures within the structure. Because of this, the following sections discuss each of the main categories in turn, though it should be realized that the location decisions in each are not independent of the others.

The commercial land use system

Central place theory was introduced in the preceding chapter as an account of regularities in the size, spacing and functions of towns. It is also relevant to the intra-urban case. A hierarchical system is the most efficient way of providing the inhabitants with the range of goods and services demanded, since the highest order functions are concentrated in the most accessible centre, the CBD, and those of lower orders are scattered among the suburbs to meet local demands only.

Development of such a system is best seen within the chronological sequence of urban growth. In a small place, all of the functions,* few of which will be duplicated among several establishments,* will be concentrated in the one centre. As the population increases, two business trends proceed: (1) *duplicating*, as more goods and services of the types already provided are demanded, and (2) *widening*, as new functions are attracted to meet the needs of the expanding market (the import replacement process). The latter process largely proceeds in the CBD, since in most cases the new functions will be serving the whole town (exceptions would be specialty stores – e.g. art shops); the former may involve a spatial diffusion process. Instead of following the pattern of previous decisions, and locating in a proven centre where competition with establishments holding an initial advantage over the market is necessary, a new, duplicating establishment may prefer to pioneer a non-central location where a spatial monopoly can be initiated over its portion of the market. (The spatial monopoly would emerge because the new establishment is closer to the required portion of the market than any of its competitors. This assumes perfect competition, of course.) The problem is to know whether such pioneering would be feasible. Theoretically, in a circular city with an even population distribution, it only becomes viable when the market can support seven establishments of any one function (Berry 1964b).[1] Because entrepreneurs do not have perfect knowledge of the situation, and most would seem to prefer playing 'follow-the-leader' (Pred and Kibel 1970), the ideal will almost certainly not be attained. As the city grows, however, the feasibility of alternative locations will become apparent, and eventually the pioneering move will be made. Of course, a premature pioneer may try a new location, and perhaps fail unless he has sufficient capital to support him until the local market expands.

Simply stated, then, the evolving form of central place provision should be that the more widespread and regular the demand for a commodity, the greater the number of establishments providing it, and the more scattered the distribution of these establishments should be. The lower the threshold for a function,* and the smaller the market needed to support one establishment,* the smaller the centre in which it is typically found. And the operation of this principle for all functions should, as in the inter-urban case, produce a hierarchical system comprising nested hinterlands.

Several factors operate in combination to prevent the ideal-type central place hierarchy developing (Johnston 1966b). One of these is population density which, as the previous section suggested, usually declines exponentially from the city centre, paralleling the similar trend for land values. Within any one distance band, however, density probably varies as a function of other characteristics of the social environment (see below). The higher the density, the larger the potential market within a given radius of any point. In the most densely-peopled inner areas, therefore, any location is likely to prove viable for establishments of the lowest order, so that such areas characteristically contain many isolated stores and street-corner clusters of two or three establishments. Locating a shop is more of a gamble in more thinly-peopled residential areas, so establishments are likely to group together; thus a broad generalization may be that the lower the population density, the larger the average size of central place.

The characteristics of an area's population are often associated with shopping centre provision there. In general, lower socio-economic status groups tend to make more shopping trips than their higher status counterparts, to make fewer purchases per trip, and to use shops as social centres as well as distribution points. Hence they are

---

[1] If the market were evenly distributed along a straight line, with one shop at the centre, a second shop could not monopolize half of the market from a non-central location. A third could obtain $\frac{1}{3}$ of the market from a location $\frac{1}{6}$ of the distance from the end, so three shops could be evenly spaced along the line. In a two-dimensional, hexagonal market there would be three such 'lines', giving seven shops in all.

better served by a dense pattern of small centres, while the higher status groups prefer the greater choice obtainable in a larger place. Comparison of two Leeds suburban populations, one with an average income of £1809 and the other of £875, showed that while the former's suburb contained 78 establishments to the latter's 65, the lower status area had many more functions present (134 compared to 101). This was because shops were more specialized in the high income area: those in the lower income suburb combined many functions under the one roof, the small centres being the equivalent of larger centres in the other area in terms of the range, though not necessarily depth or quality, of goods and services provided (Davies 1968). The functions present may also vary according to the characteristics of the hinterland, a feature which is especially so in the ethnically-oriented centres of cosmopolitan cities. Comparison of centres in 'Anglo', 'Afro-American', and 'Mexican-American' districts of Los Angeles, for example, showed that whereas food stores and eating and drinking establishments were more prevalent in the last two types of neighbourhood, professional and financial services were much more common in the first (Harries 1971).

Different socio-economic groups also tend to vary in their shopping patterns. One hundred respondents in each of the two Leeds suburbs mentioned in the previous paragraph provided information on their buying habits for fourteen goods and services (Davies 1969). Frequency of purchase varied little for nine of these: in four cases – flowers, hairdressing, shoe repairs, and, especially, banking – the higher income group shopped more frequently, while only for the purchase of cakes did the lower income people shop more often. (For the latter group, hairdressing and shoe repairs were usually performed at home.) In their choice of centre, the low income suburbanities were much more likely to use the nearest centre (81 compared to 57% in buying groceries), and to shop at chain stores. Also, if they went elsewhere it was usually to the CBD, because of their public transport orientation, whereas the higher income group shopped around much more: the latter 100 persons regularly visited 23 centres, their low income counterparts only thirteen.

The last major influence on the intra-urban central place pattern is the prevailing mode of transport at the time of the suburban development. Areas developed with linear modes – trams, buses and their equivalents – which follow fixed routes and stop frequently, tend to be characterized by continuous shopping ribbons along such routes, occasionally punctuated by clusters of high order establishments at major intersections. Where the railway was dominant, however, the pattern of suburban growth was beadlike, centred around the various stations, alongside each of which a central place usually developed. In Melbourne's southeastern suburbs, for example, the size of these centres is positively related to the age of the station; only small centres have developed alongside the interstitial stations opened at later dates (Johnston 1968a). The greater the distance between stations, the larger the associated centres tend to be. Development of other centres in these suburbs seems to have been retarded. The situation very much illustrates the 'ideal-type' described earlier (p. 62), which produces the problem of when to pioneer a new location within the suburb. In Melbourne very few entrepreneurs have been prepared to take this gamble.

Finally, in the most recent era of automobile dominance over personal spatial mobility a further form, again largely comprising a very centralized pattern, has emerged. No longer is accessibility a major constraint on shopping centre choice. Price, parking and the shopping environment have all become important attractions in the competition between centres, which are planned as units rather than haphazardly growing up. These trends have been paralleled by centralization in the retail industry, too, especially its food sector, occasioned by the inflation of wage costs relative to those of space – hence the trend to self-service, and the need for scale economies in bulk buying and selling – hence the replacement of family businesses by large companies and chain stores. Thus outer suburbs are characterized by relatively small numbers of large, planned centres, in which few functions are duplicated.

These various patterns are not mutually exclusive; most cities contain examples of them all, and they are often interdigitated, as in the outer suburbs which contain small centres to meet local needs as well as the large planned complexes. And whereas the tram, bus and train suburbs tended to be relatively separate, the newer, car-oriented facilities are being developed in most parts of many cities, causing considerable spatial reorganization within the central place system. Store vacancies, and in areas where market support is declining, wholesale blight, are one result of this. In Chicago during the 1948–58 decade, for example, the total number of establishments declined from 43,540 to 34,069 (Simmons 1964); in some areas, as many as one-quarter of all stores may have been vacant in 1960.

This reorganization of the retail system into fewer and larger centres is in part a result of time–space convergence, and so represents intra-urban centralization. Deconcentration has also occurred, not only because this centralization tends to be spread through a number of centres, at all hierarchical levels, but also through the development of business ribbons, strings of establishments along a town's more accessible and densely trafficked routes. Two types of ribbon have been recognized (Berry 1963). The first comprises those containing establishments to which special trips are made; very little of their trade is made up of impulse purchases, presumably a product of window-shopping. Plumbers and hardware supply stores, car and TV repair establishments are typical of these ribbons; they locate on main streets so as to be accessible and yet avoid the high land costs of major shopping centres. (Their need for relatively large spaces also pushes them away from the high value areas.) Bars, laundries, restaurants and drug stores are also typical. The second group of ribbons is characteristically on less important roads, their typical location being on an important suburban street, perhaps a main commuter route. In these, convenience-type establishments normally found in small shopping centres locate to capture passing trade. (Commuters getting meat for the evening meal on the way home, perhaps.)

The Macleod Trail in Calgary (Boal and Johnson 1965) is an example of the first type of ribbon, this major urban routeway containing special function, highway-oriented and general shopping goods stores. A survey found that the second group tended to obtain most of their trade from passing motorists who visit only one establishment, whereas in the case of the special function establishments, and notably the used car lots, many customers moved from one establishment to another. Shoppers in the main central place nodes along the Trail who visited several establishments usually patronized other shopping goods stores, but several combined their shopping trip with a visit to a highway-oriented function, usually a petrol station. The Trail, then, like most such major ribbons, performs a number of overlapping roles.

Finally, intra-urban central place specialization is becoming more common, as establishments become better able to draw customers from all over the city to non-central locations, though the origins of such centres often go back to the characteristics of their local hinterlands. Specialized shopping centres of the 'arts-and-crafts' type, such as Yorkville in Toronto, or the various 'markets' which originally served a small, localized ethnic group in a cosmopolitan city, exemplify such developments. Linked establishments coming together for external economies of scale provide other illustrations, such as the large medical centres with their attendant specialized services which are now in many cities, and the clusters of various types of car repair facilities.

BEHAVIOUR WITHIN THE SYSTEM

This central place system, and its continuing processes of centralization, deconcentration and specialization, provides the foci for a considerable proportion of the trips made by urban households. According to the tenets of central place theory – outlined in the preceding chapter

– there should be considerable spatial order, based on the minimization of transport costs principle, in this trip pattern.

Shopping for convenience goods, those items bought both frequently and regularly, to a large extent conforms to the 'nearest centre' principle. This is presumably because the establishments retailing these much more closely approximate the assumed perfect competition between each other than do those selling comparison or shopping goods, those purchased with lesser frequency and regularity, for which there is usually a much greater choice, in price, quality and style. Of a sample of 495 Christchurch, New Zealand respondents, for example, only about ten per cent generally purchased groceries and meat from more than one centre, and the nearest centre was most frequently chosen (Clark 1968; this survey was conducted before large supermarkets were common in the city). Those who did not patronize the nearest alternative, however, tended to move much further afield, rather than choose another centre in the local neighbourhood. For comparison goods – clothing, furniture and household appliances – most people used only one centre, the CBD, by-passing closer alternative centres and apparently being almost completely unconstrained by distance (Clark and Rushton 1970). (Christchurch is a relatively small city of about 250,000 persons only, however; a larger centre may display a more distance-constrained pattern as people select among major suburban, probably planned, centres.)

Many reasons can be suggested why some people do not shop at their nearest centre. Multi-purpose trips, for example, may lead shoppers to get convenience goods at a distant centre where they also purchase higher order commodities. Others will do their shopping as part of another trip; many working housewives, for example, shop at the nearest centre to their place of employment. The spatial system is thus still being used rationally, but the home is not the origin of the trip on which costs, and probably more importantly, time, have to be minimized. Moreover, attitudes to shopping vary. Some people will cross a large city for a day's outing at a large, planned shopping centre; some eagerly scan the papers each week to see where the best 'specials' can be obtained; others consider the task a chore, to be finished as soon as possible. Finally, few shop types are perfectly substitutable, even in the grocery trade, because of different selling methods, and customers, with their ever-widening indifference to the frictions of distance, are prepared to travel to the type they prefer. Fewer housewives in 'developed' countries are now constrained to the shops within pram-pushing distance, though this varies from place to place, perhaps as a consequence of different car ownership rates.

Part of the complexity in shopping behaviour patterns stems, at any one point in time, from the interdigitation of central place systems developed under different technologies. The existing system of un-planned centres must readjust to the new competition from large planned centres. In Wellington (N Z), the CBD is becoming more dependent on local trade, especially that provided by its workforce, and less on its ability to attract suburban housewives. When a new planned centre opens, neighbouring unplanned centres – of all sizes – find their hinterlands attenuated, as people who live some distance from opportunities transfer even their convenience goods shopping to the new facilities (Johnston and Rimmer 1969). Large supermarkets placed in existing small centres also tend to complicate patterns. Customers will travel relatively long distances to patronize the super-market but will probably not visit the small centre alongside it, pre-ferring to make necessary purchases at the small centre next to their home which they have always patronized. In this way multi-purpose trips become multi-destination trips also.

With rapidly changing technologies in the supply sector, consider-able flux in consumer demands, and increasing spatial mobility for a growing proportion of the population, it is unlikely that any condition close to equilibrium will be reached with intra-urban central place systems, especially in larger towns. Instead, developments in any of the three spheres mentioned, which usually affect different areas or different population groups at varying times or rates, will ensure a continual process of change, and thereby make it difficult to elicit the spatial order in the system's operation.

INSIDE THE CENTRAL PLACE

Just as the individual urban place is a sub-system of the inter-urban system, so each individual central place can also be studied as a full spatial system, as an ordered pattern of interacting land uses. Thus a pioneering study of the CBD's internal structure (Murphy, Vance and Epstein 1955) showed that the core area – housing the retail, service, financial and office functions – could be divided into four zones around the most accessible point, the peak land value. Retail businesses, as suggested in fig. 3.1B, dominated the inner zone, because of their reliance on pedestrian traffic – variety and clothing stores were most common there. With increasing distance from the centre, shops were replaced by offices as the dominant function, firstly by those offices needing contact with the general public, and then by headquarter and finance offices. Within each zone, however, were clusters of linked uses. Among the retail uses in Austra-lian CBDs not in the inner zone, there are tendencies for separate clusters of furniture and hardware stores. In the office zone, insurance companies tend to occupy the best sites (with the newest buildings), shipping offices are further out, lawyers cluster around the courts, and doctors often group in rows of once-fashionable town-houses (Scott 1959). Other functions, such as theatres and restaurants, also cluster, but others still serve a dispersed clientele throughout the CBD; establishments such as barbers, banks and bars tend to be scattered throughout the area.

On the fringe of the CBD, land use tends to be less intensive and more externally-oriented. Wholesaling, transport, and some industry typify this outer area. Many of these functions are being moved to the suburbs as part of the deconcentration process, however, and with the already noted decline of the CBD's retail function, this leaves the office as increasingly the predominant city centre land use. In large cities such as London, a complex spatial system of office areas has emerged, and block-by-block analysis has identified five main type areas (Goddard 1968). Apart from the money market area of the financial core, these are: the commodity trading, risk insurance, and shipping quarter; the capital and investment finance area; the publishing and professional services group; and the textile trading companies. Grouping of these like uses into various office quarters presumably results from a need for rapid personal communications. Analysis of taxi flows within central London suggests such a pattern. Recently, a more detailed study involved a sample of businessmen recording all of their business meetings during a three-day period. One-third of the meetings involved leaving their offices, and one-third of these involved less than a ten minutes' walk (Goddard 1971). This seems to confirm that location and inter-office linkages are inter-related.

In recent years, however, the intense centralization of offices in London has relaxed slightly, another indication of the deconcentration process. Some offices have moved to provincial centres; more have gone to outer suburban nodes, such as Croydon and Reading. Most remain within central London, but no longer apparently need to be quite so clustered. In 1918, for example, publishing was concentrated around St Paul's and Covent Garden, while by 1938 a third cluster close to the British Museum was flourishing. Since then, there has been some westward migration, where most of the newer firms have established premises. Offices are now less tied to particular parts of the city; 'improved communications enable certain firms to maintain essential contacts over a longer distance. They therefore feel free to locate anywhere within a relatively large central area ... that satisfies other conditions such as an attractive working environment and a prestige address' (Goddard 1967, 283–4).

Migration of functions within the central area is not unusual, and indeed the core of the CBD itself often migrates over time. Separate zones of assimilation and discard, or advance and decay, were noted in early studies of American CBDs. The former has been renamed the sector of active assimilation (Griffin and Preston 1966); it tends to move the CBD towards the higher status residential areas and away from the wholesaling area; its characteristic functions include specialty shops and new office and hotel blocks. On the opposite side

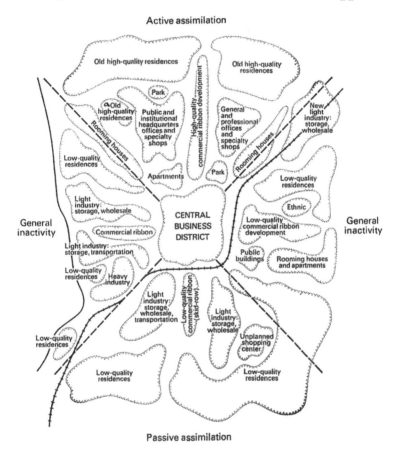

fig. 3.2 *Generalized pattern of land uses in the transition zone around the* CBD. *Source: Griffin and Preston* (1966).

of the CBD, this sector is complemented by one of passive assimilation, or decay (fig. 3.2). Here industry and dilapidated housing are often inter-mixed, along with many cheap bars and cinemas, credit stores, and, where they still exist, pawn-shops. And finally, these two sectors tend to be separated by areas of general inactivity, where change is but slight; often industry is well entrenched in such areas.

Although the CBD is the largest and most intense collection of commercial land uses, it is not alone in having a clear-cut spatial system. Garner (1966) has adapted central place principles to the intra-centre situation, and shown that large suburban nodes should

also be characterized by a zonal structure, with a negative relationship between an establishment's threshold and its distance from the
peak land value. Certain types of shop also tend to locate in groups,
because of their complementary functions. In Liverpool, grocers,
greengrocers, butchers, confectioners and chemists tend to cluster
with each other, as do sweetshops and cinemas, shoe and dress shops,
record stores and milk bars, fish and chip shops and off-licences (Parker
1962); others, such as pet food stores and butchers, are mutually
repellent, and are rarely found on adjacent sites. Surveys in Australasia
show that shoppers tend to patronize such groups of stores, rather
than move about at random within the centre (Johnston and Kissling
1971). Transport costs reduction and efficiency in movement thus
seem to influence spatial behaviour at even these micro-scales.

The commercial systems discussed here are only certain levels of a
nesting hierarchy of spatial structures that proceeds downwards from
the international system to the individual room. Each shop or office
is itself a spatial system, with its specialized parts between which
people, goods, and communications move. But micro-scale investigations of these systems have been avoided by geographers, who focus
on wider areas in which, despite much complexity, relatively simple
principles of spatial order seem to influence much of the behaviour.

Industries in the city

The 'classical' model of urban spatial structure, propounded by Burgess
in the 1920s, placed the industrial districts in a collar around the
CBD. This was a major weakness in the model, probed by a leading
critic (Davie 1938), and accepted by a later advocate (Quinn 1940).
Nevertheless, many industries in many towns were concentrated in a
circumferential belt around the business area. Exceptions were smaller
towns whose growth depended entirely on their new industry. Usually
this was a large plant which located at will and often, as in railway
towns such as Swindon, led to a major reorientation of the urban
morphology. With technological developments, however, these
'exceptional' locations have become the rule.

Changing industrial locations are illustrated in Birmingham,
England, where a three-zone structure represented the general pattern
before 1939 (Wise 1949), much of which remains today. Spatially-
extensive, 'modern' plants, as in the automobile and aeroplane
industries, typify the outer ring of post-1918 suburbs, being preceded
in the next zone by the factories of the late nineteenth/early twentieth-
century Birmingham trades – bicycle, machine tool and electrical
apparatus manufacture, for example. Finally, in a belt around the
CBD are the quarters of the long-established trades, notably those

involved with gun and with jewellery manufacture. These contain many small workshops, existing in cramped quarters and linked together in long processing chains for the production of the final article. Initially, workplace and home were on the same premises, and clustering was needed because of the many interactions involved in the production process. Neither factor is as important now, but the quarters remain.

Different locational requirements account for this pattern. Several industries, for example, comprise a large number of small establishments, with a myriad interactions needing rapid contact among them. The CBD collar is their typical location, the availability of buildings – rarely custom-built for the purpose – being a further prime locational determinant. Analysis of London's industries since 1861 showed that these features have characterized printing, clothing, furniture and precious metal fabricating (Hall 1962). All four were concentrated in the inner belt in 1861; the first three still are. In the clothing trade, access to market was important, especially for bespoke tailoring, 'because of the importance of individual fit and the capricious and unpredictable trends of style' (Hall 1962, 52). For the wholesale trade, small specialist workshops prevailed over factory organization (in London, at least), and even where these were involved in the ready-made sector, market access for information on style changes was vital. Finally, the many different types of workshop between which the cloth moved were attracted by the availability of a local labour force, especially women and the Jewish population who lived in the area. For the other industries, too, 'in one form or another, nearness to the market has been the original locating factor' (Hall 1962, 115). With the printing trade, for example, speed in returning the printed material to the source of copy was often imperative, as it is still for many legal and administrative functions, and in newspaper production.

Factory clustering is also a feature of the suburban zones; indeed, town planning legislation has made this virtually obligatory in many countries during the last three decades, though the clustering is a natural process which predates the planning ideal of land use separation. One such area in the inner suburban ring of Greater London, Park Royal in western Middlesex, occupies the site of a World War I munitions complex. Relatively small firms (average number of employees in 1952, 105), especially in the vehicle, metal manufacture and general engineering trades, were attracted by its relatively cheap, ready-built factories. Several such clusters exist in Greater London, and in other cities too, some dominated by particular industries which grew rapidly during the inter-war years. Food processing is one of these, chemical manufacture another.

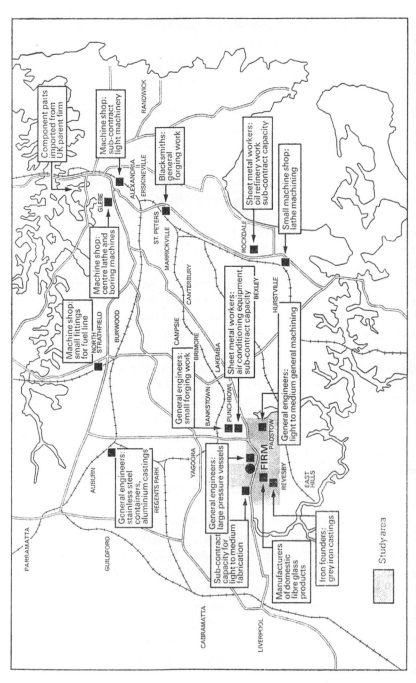

fig. 3.3 The linkages of a manufacturer of earth-moving equipment whose factory is located in suburban Sydney. Source: Logan (1964).

The availability of space, labour and buildings have all been mentioned as factors influencing the location decision, though the last of these is relatively unimportant to the largest firms. Access to the market is often critical, too. Many firms moving to, or opening branches in London's outer suburban zone quoted access to the capital city as a major influence on their choice of site (Keeble 1968). Indeed, half of them would not have abandoned more central locations, either partially or totally, if they had had to go further into the provinces: over half of the firms interviewed shipped over 40% of their production within London and the Home Counties.

External or agglomeration economies are also relevant for these outer zone industrialists, many of whom prefer to rely for their supplies on sub-contractors, who must be accessible. Large suburban industrial estates often generate the development of such contracting firms, both in specialized manufacturing tasks and in the provision of services, such as transport. In such an estate at Bankstown, Sydney, over 70% of all the firms interviewed used subcontractors, 80% of whom were located on the estate itself (Logan 1966). Local firms were preferred because of cheaper and more rapid service, plus a greater ease of inspection, but often the manufacturer must go further afield, resulting in a pattern of linkages like that shown for a Bankstown manufacturer of earth-moving equipment in fig. 3.3 (Logan 1964). It should be noted, however, that many of Logan's respondents found this system at least partly unsatisfactory and hoped to replace the subcontractors by expanding their own production when capital was available. Such replacement of external by internal economies is generally available only to large firms, however, and though the overall trend is to larger concerns, it is unlikely to remove completely the need for the type of spatial system described here.

The relatively large firms located at Bankstown represent only one type of suburban factory. At another node, also of relatively large plants but closer to the city centre and with much less space, external economies were much less important, since only 25% of the firms used subcontractors. Perhaps this pattern is typical of an older area, where external economies are replaced by internal ones, and may be the 'end-state' for the Bankstown area. Finally, an area of small firms in the northern suburbs had only one-third of its plants using subcontractors, most of them in other parts of Sydney. It is characterized by industries whose products are small and light, such as cosmetics, which serve a national market, and which require relatively few linkages with the other industries.

According to these research findings the importance of external linkages varies by firm size and type, and location decisions are at least in part related to these. The linkages themselves also vary, both in complexity and in nature. Some firms may interact with only one or two sources for inputs or destinations for outputs; others may interact with many. For many factories the linkages involve the movement of goods; for some it is the flow of information that is vital. Thus a typology of linkages for firms can be devised, as shown in fig. 3.4 (Wood 1969), though it should be recognized that several firms may fall into more than one category, if both material and information flows are critical to them.

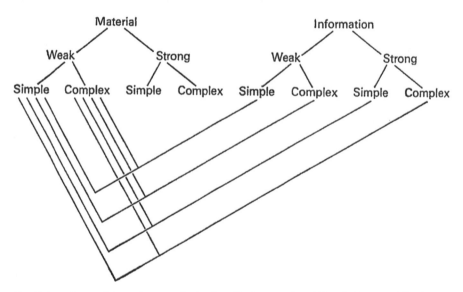

fig. 3.4 *A typology of manufacturing linkages after Wood* (1969). *The lower portion of the diagram shows only a selection of the possible combinations when both types of linkage are considered.*

THE CHANGING PATTERN

Despite this concentration on the role of linkages, the intra-urban industrial system, like the commercial system, is not now as dependent on close spatial propinquity between plants as may have been true in the past. Again, time-space convergence is a major reason for this changing spatial order. Wider scatter is now possible. For Birmingham's jewellery trade, Wise (1949, 63) wrote that 'whereas concentration remains exceedingly high in the heart of the quarter some dispersion has been characteristic on the fringes and particularly among those firms producing cheaper quality goods'. And in the outer suburbs of north-west London, Keeble (1969) has emphasized the importance of a wider spatial matrix than the local, especially for the larger firms. The latter, it seems, can move distances of up to 100 miles and still retain satisfactory contact with their various supply and demand points, though they much prefer to remain in the same metropolitan sector.

Other processes of change in the industrial system have clear spatial co-ordinates too. Over time, old industries may decline; they may be successfully replaced so that the town retains its economic viability. Within existing industries, firms close and others open. The sum of these two types of change has been mapped for Merseyside during the 1949–1959 decade, using a kilometre square grid. A clear set of spatial patterns emerged (Lloyd 1965). First, an entry ratio was computed, as the ratio between the number of plants opened during the decade and all plants existing there at some time in the period. Not surprisingly, this ratio tended to be highest in the new, outer suburban areas (fig. 3.5). Secondly an exit ratio was computed, with the same denominator as before, and with the number of closures as numerator. A tripolar spatial pattern was displayed for this index (fig. 3.5), with high proportions of closures not only in the centre, where they might have been anticipated, but also in some of the newer suburban areas. The latter mainly comprised small plants, perhaps sub-contractors attracted to the developing nodes which proved to be nonviable locations for them. Finally, a survival ratio, for which entries surviving to the end of the decade formed the numerator, showed that turnover was by far the greatest in the inner areas. Up to 60% of all new plants survived in the outer suburbs – over 35% in all of them – but in the conurbation's inner district only 18–20% lasted through the time period.

Similar findings are reported from an investigation of Melbourne's industrial system over the 1950–65 period (Rimmer 1969). Of the factories present in one inner suburb at the first date, only 20% remained in the same hands fifteen years later, with a further 7% having a different owner but still in the same industry. Yet during that time, only five units were demolished and twenty new ones built. This suggests a further role for the inner areas, that of a 'seedbed' for new firms. Small, cheap premises available for rental allow potential factory owners to establish their own businesses. Many of them fail, to be replaced by further aspirants, often in a different industry. Others succeed and then move out to suburban locations, where space, and often labour, is more plentiful, and premises may be bought rather than rented. Within Melbourne, as in London, the majority of such movers remain within the same city sector, thereby retaining access to their suppliers, their clients and their workers.

There would appear to be several dimensions to the patterns of industries within the city, therefore, with access requirements, plant size, and age and stability of the firm among the more important of these. The result of the interactions of this variety of influences is a

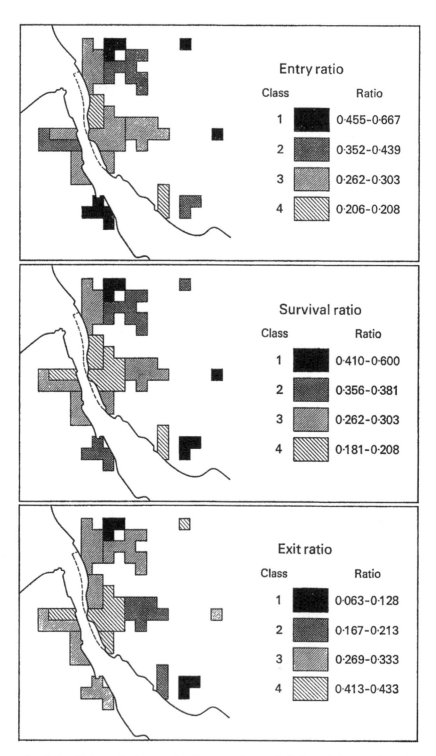

fig. 3.5 *Rates of change in the numbers of industrial premises by kilometre square in Merseyside. After Lloyd (1965).*

complex spatial system about which researchers have found it very difficult to generalize, for industries cannot always be easily categorized into locational types. Some, such as those which are waterfront-based, are easily classified; many are not (Pred 1964b). Rimmer's (1969) analysis of 169 industrial types in the Melbourne metropolitan area suggests some general pattern, comprising two basic groups, each divided into three subgroups. But this classification, and others like it, suffers several defects. By concentrating on the product, it tends to ignore the important input–output relationships which, together with cumulative inertia in many cases, are dominant influences on industrial locations. Division into 169 industrial categories avoids some of these problems, but it does not allow for quality differences between plants making the same products. Furthermore, Melbourne's industries are perhaps much more oriented to that city's internal markets than may be the case for centres which are part of a denser network of large places, so patterns there may be peculiar to a certain type of city only. In general it may prove impossible to develop classifications of locational types based on products, so that attempts to discern order in this element of the intra-urban system have to resort to other criteria for classifications, such as the linkage typology outlined in fig. 3.4.

The intra-urban residential pattern

The land-value model outlined at the beginning of this chapter, and the subsequent discussions of commercial and industrial components of the intra-urban system, assign residential land uses to the residual areas, those not required for other uses. But these housing districts, in almost every town the largest consumers of space, merit detailed consideration because they differ considerably among themselves. This section outlines the nature of these variations, and their characteristic spatial forms.

THREE DIMENSIONS

Three fundamental questions underlie the study of intra-urban residential patterns: do different population groups tend to occupy separate residential areas, and if so why?; how many such basic dimensions of this differentiation exist?; do the different groups tend to occupy certain type locations relative to other land uses in the system?

The existence of residential separation has for long been accepted, and most research in recent decades has focused on the second question: how many dimensions to the patterning? The general consensus from work in a variety of social environments – though most of those examined have a basic Anglo-Saxon culture – is that there are three: socio-economic status, family status, and, where relevant, ethnic status. These have been largely determined from a wide array of empirical studies many of which derive from the work of Shevky and Bell (1955), who first suggested the three dimensions on the basis of their investigations in Los Angeles and San Francisco. Their major hypothesis, which underlies one of the main themes of the present book, was that with increasing levels of economic development, societies are characterized by three trends: occupational differentiation which is translated into social stratification; a wider range of life styles, especially with regard to the family's role in society; and population redistribution, which brings together people of various ethnic and cultural backgrounds into urban places. Shevky and Bell derived indices of these three processes for intra-urban census tracts, and demonstrated that the population of the city was residentially separated according to the three dimensions. No explanations were offered as to why such a residential mosaic should result from the postulated social trends, however. (For a fuller review of this topic, and many of the others dealt with in this section, see Johnston 1971a and Timms 1971.)

Other authors have provided theoretical underpinnings for the patterns which Shevky and Bell observed, although not all of these have been fully substantiated. In general, it might be said that groups separate from each other in order to maintain their identity and to reduce contact with other groups. For some, the location of the group's neighbourhood relative to other parts of the city may be immaterial. For many it is not, however, so the system of residential differentiation reflects group attitudes to space and accessibility and is a spatially ordered system which facilitates the desired contact patterns, although there is usually inequality in the degree to which the desires of various groups are met.

In a large city, socio-economic status display is difficult, since each resident can know only a small proportion of his fellows. To transmit one's status in general terms, housing and, in particular, one's address are widely accepted as a symbol of position in society. Unlike many other symbols, address is not usually ephemeral; the characteristics of an area's social environment often remain constant for several decades, even though the residents may change several times over, and these characteristics become widely known, through such generally accepted indices as property values, neighbourhood appearance, and dwelling styles. There seems to be general agreement within cities on any district's social position, and this is to a large extent paralleled by residents' aspirations for living in such areas; the higher a district's status, the more desirable it is as a living area for most people (Johnston 1972c). The desirability is reflected in the prices people are prepared to pay to live in such areas, and as social stratification is characterized by variations in rewards (incomes), this ensures that only certain groups can afford to live in the various neighbourhoods.

The model of the residential location process described in the last paragraph assumes that all socio-economic groups aspire to live in the most desirable area they can afford. Evidence is inconclusive as to whether this is so, or whether in fact only the higher-income groups have such aspirations and the remainder of the population must go to the residual areas. Various studies suggest the general validity of the model, although there are many deviations from it, not only the well-to-do who choose to 'slum' it in poor neighbourhoods but also those relatively affluent sections of the middle class who are satisfied with their homes and neighbourhoods and do not move to anything 'better', even if they can afford to do so. But in Chicago, lower level white-collar workers are prepared to expend more of their income on housing than are the, often better paid, skilled blue-collar workers, so that they may live in the 'right' areas (Duncan and Duncan 1955); that they may have local contacts – in the street, church, and school – with their social peers and superiors. And even in government (national, regional or local) housing areas, as in Britain, a grading of neighbourhoods often becomes apparent (Elias and Scotson 1965) and households apply to move into the more 'respectable' districts which they feel befit their status.

Many of the hypotheses presented here, and most are little more than that, suggest that distance has an important influence on social interaction, and that the choice of residence is an explicit reaction to this. Close correlations between distance and the distributions of social contacts, of friendships, and of marriage partner choices have all been shown, yet often the studies are unable to indicate whether distance is a causative factor or merely a joint effect of the desire to interact with one's peers and the existing pattern of residential separation. Careful research, as in Ramsøy's (1966) study of marriages in Oslo, has been able to demonstrate an independent role for distance as an influence on, if not a determinant of, social interaction patterns. It has yet to be shown, however, whether this is merely a result of residential location decisions made on other criteria, or whether people are indeed adjusting to this known spatial order.

Spatial separation of various groups in terms of their life-style orientation probably results from a two-stage process. First, there is a basic difference between the family-oriented and the non-family-oriented households in their dwelling requirements. Space, both internal and external, is usually in the forefront of the former's demands, because of both household size and composition and the total amount of time spent in the home. For the non-family-oriented, usually smaller, household, space may be much less of a desirable commodity, as less time may be spent at home. Security of tenure is another desideratum of the family group, the house being considered a major investment and security against future inflation or hardship. Ownership is thus considered very desirable, though societies vary in the degree of social, and political, pressure exerted to this end. Non-family households, on the other hand, will usually opt against ownership, not only because they have less need to consider long-term security, but also because this form of tenure restricts their mobility and may involve expenditure, of time as well as of money, in upkeep of their possession. Finally, the two groups may differ in their accessibility requirements. For the family, a whole range of trips may be made each week to a variety of destinations – schools, shopping centres, etc. – only one of which will be a member's place of work. The latter journey may be relatively insignificant in the residential location decision, especially if the commuter has a car or good public transit available to him, while his wife must push a pram. To many households, neighbourhood school quality is a very important criterion. For the non-family-oriented household, however, the journey to work may form a much larger portion of the total trips undertaken, especially if more than one member of the household works, and other major trips – perhaps to places of recreation or entertainment – may be to the same general area, the CBD.

That the above environmental requirements lead, in combination, to a spatial pattern of different types of residential area depends on different parts of the city being best suited to meet certain needs. With the space requirement, for example, the land value model suggests that residential land some distance from the main commercial nodes is more desirable than that closer in since it is much cheaper. The more valuable land provides a better return if it is more densely occupied and so is more likely to be developed in flats or apartments for the non-family-oriented. All land offers a higher return from higher density use, of course, so it is other requirements, such as that of access, which lead to a more central location for the multi-household dwelling units. (And with time–space convergence, this is becoming less crucial, as will be pointed out later.)

The second stage in the development of residential differentiation along the family status dimension involves a further pattern within the family-oriented areas. In this, households are distributed among the various areas according to their age structures, or stage in the family cycle. To a large degree this comes about because families obtain homes which meet their requirements and then may not move again, certainly until the family is contracting in size or dissolving. Since most districts are developed as units, they tend to be occupied by people of the same age (as well as income) who were in the house-purchasing market at that time. Thus separation by age group often ensues, and some researchers suggest that this is continued by others wanting to live among their generational as well as their social peers. At the later stages of the family cycle, developers often cater specifically for requirements with housing areas for retired persons.

Residential differentiation along the third of Shevky-Bell's dimensions – ethnic status – is a spatial consequence, and often reinforcement, of the assimilation/integration process by which immigrants enter the urban society. Many of these people are of low socio-economic status, but this is not the only reason for their spatial separation (Darroch and Marston 1972). Often they are the recipients of prejudice from the host society, and are discriminated against in the housing market, so that they are unable to exercise a free choice in where to live. 'Visible' minorities are usually treated in this way, those of alien culture, language, customs, skin-colour and, in some situations like that in Northern Ireland at present, religion too. In many cases, this discrimination may be ephemeral and members of the group later welcomed into the wider society and its residential areas. In others, it is not, as seems to be the case with the Negro in United States cities, where discrimination is leading to permanent 'ghettoisation' (Taeuber 1968), which also results in the development of a residential mosaic within the ghetto (Meyer 1971).

Housing market discrimination may force a group to become very inward-looking, so that it decides of itself to maintain its social distance by spatially isolating itself. One result of such action may be the continuation of ghettos long after social and spatial assimilation would have been accepted; this could be so for many Jewish populations. In many other cases, however, self-separation may result without any overt discrimination, but merely as the wish of the persons involved. Many migrants in an alien culture cluster together for mutual reinforcement, to continue the cultural patterns of their place of origin and to use culture-specific facilities, such as churches using a certain language. This is particularly so for many aliens who move in chain migrations, in which the feedback of information and money from pioneer migrants to their home areas produces further migration. Many very tight clusters of Southern Europeans and of Polynesians in Australian and New Zealand cities have been produced by such chains; families from the same village live next door to each other in a city 12,000 miles away (Burnley 1972). Such groups are as much segregated among themselves as from their host society. Over time, however, there is usually some leakage as people leave the urban village and join the wider society. But the cluster generally continues for as long as immigration does.

FACTORIAL ECOLOGIES AND THE DEVELOPING RESIDENTIAL MOSAIC

Existence of these basic dimensions of residential differentiation has been confirmed for a large number of cities. Some early studies, such as Shevky and Bell's, used simple indices to demonstrate their findings (as also did the Chicago studies of Duncan and Duncan 1955, and Duncan and Lieberson 1959). The majority, however, have used the methodology of factorial ecology which applies the techniques of principal components or factor analysis* to data matrices in which various measures of population and housing characteristics form the variables and small areas within the city (census tracts) form the observations. The first of these studies used the seven indices of the Shevky–Bell method to demonstrate the necessity for three separate, though related, dimensions to account for the variation in census tract characteristics. With the advent of large computers and the greater availability of data, factorial ecologies conducted since 1960 have used much larger data sets (comprehensive reviews of this literature are given in Rees 1971, 1972).

The 'developed' countries of the United States, Canada, Australia, New Zealand and Britain predominate in the bibliographies of this set of studies. In almost every case, they show that socio-economic and family status are major dimensions of differentiation (though, because of data deficiencies and the large public housing areas in most towns, the British results do not always completely conform to this). Other dimensions often appear, but they can be classified into one of two types:

(1) The splitting of one of the major dimensions. Socio-economic status, for example, may be represented by two dimensions, one reflecting an income-dominated pattern, the other based on education/ occupation (thereby suggesting differences between 'wealthy' and 'fashionable' residential areas). Sometimes these splits reflect the data input or the local situation (ethnic status may comprise several dimensions in a cosmopolitan city). Others indicate the high level of generalization which the method provides, as a simplification of a more complex reality not yet fully understood.

(2) New dimensions, which do not invalidate the initial model since they are often based on variable sets not included by Shevky and Bell. Population mobility, new suburban development and 'Skid Row' are such additional dimensions to the total pattern.

Finally, of course, ethnic status will be absent from a city with no alien minority group.

There is considerable evidence that three major sets of factors produce the urban residential mosaic, therefore, though research in depth is far from complete on this topic (as shown by the papers in Berry 1971a, 1972). Although the results discussed so far relate to only one 'culture realm' there is evidence from a variety of other environments – Indian, South American, Korean, Italian and Scandinavian (Rees 1971, 1972) – that the same basic dimensions exist. They may, of course, be represented by different indices, as with single versus multiple dwelling units in Boston compared with Helsinki (Sweetser 1965), but the underlying causes do not vary. Other studies in 'modernizing' countries, however, show that a smaller number of dimensions is necessary there. In Cairo, for example, life style choice is apparently a function of socio-economic status, and not independent of it, so a single major dimension suffices (Abu-Lughod 1969); in Poona, the main characteristic differentiating areas over the last 150 years has been the origins within India of their residents (Berry and Spodek 1971).

The major axes of residential differentiation would seem to change with economic development, in the following way (McElrath 1968). At the lowest levels, urban populations are segregated according to their 'tribal' origins, and development from this position sees that dimension replaced by one of socio-economic status. Gradually, family status emerges independent of this first dimension, until the position already noted above is attained, when it is a completely separate major factor. Such a sequence can only be inferred from cross-sectional comparisons, as in Schwirian's (1972) analysis of the three main Puerto Rican urban areas, and further work is definitely required on this topic.

WHERE DO THEY LIVE?

Factorial ecologies and like studies illustrate the existence of differentiation within the residential system, but further analysis is necessary to enquire whether the different groups tend to occupy certain type locations within that structure. The earliest model referred to socio-economic status patterns and suggested a gradient of increasing status with distance from the city centre (Burgess 1924; Johnston 1971a). This was later modified to a series of sectors of relatively homogeneous status aligned along main transport routes, but still retaining a similar zonal pattern within most sectors (Hoyt 1939). Since they are influenced in their residential location by their low socio-economic status, most ethnic clusters are found in the inner zones close to the CBD, although their 'ghettos' may extend considerable distances out along certain axes. Finally, no formal model of family status patterns has been propounded, usually a zonal pattern is assumed, with non-family areas closest to the centre, and, within the family-oriented areas, decreasing average ages with increasing distance from the CBD.

The validity of these three patterns has been tested several times, with the usual conclusions that socio-economic status patterns are sectoral in form, family status patterns are zonal, and those of ethnic status are clustered (Berry and Rees 1969). The ideal pattern is thus an overlay of the three (fig. 3.6) though that simple system is always somewhat distorted in any one particular site. In any case, it is often claimed that this pattern is peculiar to the North American situation, and that in other countries preferences for locations vary, so the pattern does too. Schnore (1965), however, has postulated a general

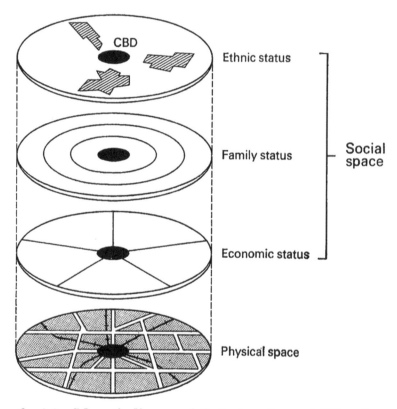

fig. 3.6 *Schematic diagram of the basic patterns of differentia-tion among residential areas in a 'Western' city. Source: Murdie* (1969).

model in which with modernization a city changes from having its highest status residents in the central city and the lowest status on the periphery, through an intermediate stage with high and low in the centre and intermediate on the periphery, to a final one with low in the centre and high on the periphery. He claims to recognize this process proceeding in Latin America and, through cross-sectional analyses, within the United States. Alternative evidence, however, suggests that in both areas, and in Australasia too, the more typical situation is for the highest status suburbs to be located in an intermediate zone (in most sectors and not only that which is overall of highest status) and for the outer zone to be occupied by the rapidly expanding middle-classes, albeit in dwellings of very different absolute standards in inter-country comparisons (Johnston 1972d). Very high status 'ex-urban' villages may form an outer zone of some cities, particularly very large ones such as London.

A FOURTH DIMENSION: THE JOURNEY TO WORK

A further factor which influences many residential location decisions, especially in larger cities, is location with relation to workplace. Other facilities of comparable standard, such as schools, shops and homes, may be available in several locations, so the choice may be in terms of accessibility to an employment area. For people at the two extremes of the income continuum, such choice may be very limited and long commuting trips may be necessary because of their fixed home locations; for others, however, it may be possible to trade accessibility against other features, such as space (Dahms 1971). Among a sample of nearly 700 householders in Newcastle, New South Wales, for example, about 14% quoted nearness to workplace as a major determinant of residential choice, and another 18% cited it as a minor influence (Daly 1968).

The generally observed pattern for commuting is that workers tend to cluster their homes around their places of employment, especially when the latter occupy suburban locations. Such clustering is particularly true of most blue-collar and poorly paid workers, and also of female employees. It has been suggested that accessibility levels are now such in American metropolitan areas that commuting is no longer a constraint on residential choice (Stegman 1969). This would seem to be an exaggeration of the situation, although in general people obviously have much wider spatial horizons than urban residents of even fifty years ago, and may be indifferent to increasing commuting trips – measured in time or cost units – over certain lengths.

THE CHANGING SPATIAL SYSTEM

Over time, the ways by which population groups are sorted into districts may remain very stable, but the characteristics of particular areas of the city may vary. On the socio-economic axis, for example, the general expectation is that an area's relative status will decline over time, as its dwellings age and deteriorate and the residents move to better homes. This process, termed filtering, is usually initiated by the construction of new, relatively expensive homes which high status households occupy, their former homes being made available to lower status groups – who are thus able to 'filter up' the housing scale. Occasionally, as indicated on p. 86, such new construction and filtering may be generated by the expansion of ethnic status areas within the inner zones, which generally elongate along a sectoral axis. With regard to areal variations in family status, the major process is that of population ageing. Expansion of the inner, non-family-oriented zone may occasion more noticeable change in an area's population character-istics as houses are either subdivided into, or demolished to be re-placed by, flats and apartments.

These three basic patterns of change have been identified many times, including a number of factorial ecologies of change data, where the variables are the differences in, say, the proportion of high income earners between two dates (Murdie 1969). Filtering, socio-economic status change, ageing, life style changes, and ethnic 'invasion' are all proceeding at the same time, though usually at different rates. The last two are usually the most explosive. One immigrant group may 'take-over' a whole neighbourhood very rapidly if its demand for housing is great enough. In many cases, as with the expansion of Negro ghettos in American cities, such invasion is often restrained for a period, but once it begins it then proceeds very rapidly, as the former residents may panic and sell their properties because they have no desire to live in a 'mixed' neighbourhood. On the family status dimension, ageing of an area's population is a relatively slow process, but the transition from family to non-family-oriented may not take long. Once the breach has been made and a few homes subdivided, or a few apartment blocks built, the whole neighbourhood may quickly succumb, perhaps also because people do not wish to live in a 'mixed' situation. The same may also be true with the socio-economic status decline of a neighbourhood but, unless this is associated with one of the two processes just described, it is usually very gradual. Indeed, as noted above, the characteristics of many areas on this dimension may remain the same for generations (Johnston 1966c; 1973).

Changes in spatial patterns are also sometimes generated by major changes in the social structure, and these are apparently taking place in the family status dimension in several countries at present. The nature of the family cycle has altered radically in recent years. Pre-viously, women continued to have children throughout most of their fertile years; now, a majority have completed child-bearing by the age of thirty (Gilson 1970). Together with the average younger age at marriage and older age at death, this means that couples spend much larger proportions of their married lives without children at home. The traditional spacious single-family-house-with-garden, which typifies 'British' societies at least, no longer meets dwelling require-ments for much of the family cycle. The apartment is becoming much more popular, therefore, and further dwelling forms are introduced or adapted for such households.

Not only is family structure changing, but the divisions between family-oriented and non-family-oriented households are becoming less clear in many cities. No longer do many wives have to choose between a career or children. Many have both, employ trained persons to care for the children during the daytime, and prefer to raise their families in apartments rather than in houses. Thus custom-built apartments, rather than subdivided houses, now dominate in many countries, and they are being built in a wide variety of intra-urban locations. Among six Canadian Prairie metropolitan areas, for example, apartments ranged from 14 to 29% of all housing starts in 1960; in 1969 the range was 47–68%. The blocks have been getting bigger too, with only 17% containing more than 100 units before 1961, but 63% during 1965–8. And finally, they are being built throughout the city: before 1961, 90% of Winnipeg's apartments were built within $3\frac{1}{2}$ miles of the CBD; during 1965–8, 41% were built miles from that point (Nader 1971). Previously, it was pointed out that family status only became a major dimension of residential differentiation at a certain stage of development; perhaps the present trends indicate that at a later stage its significance begins to decline.

BEHAVIOUR WITHIN THE SYSTEM

As the spatial pattern changes, as fashionable areas decline, ghettos expand, or homes are converted to flats, people move. In all three of the above cases, most people would move towards the urban periphery, and indeed this is a major pattern in the distribution of migration directions within the city. Such moves are only a very small proportion of the total amount of spatial mobility within most cities. Urban residents move frequently, both between and within cities. Usually, the characteristics of those households moving into and out of a dwelling differ but slightly, so that the peculiar features of an area may remain despite the turnover of residents (Simmons 1968, gives an introduction to this topic).

The majority of moves are made to adjust people's housing provision to their requirements. Some may change address to obtain a home fitted to a new social and financial position, either better or worse than that preceding it, or to leave a neighbourhood becoming rapidly filled by 'undesirable' elements. But the majority, according to several surveys, are made because the previous dwelling no longer fits the space requirements of the household. Because of this, most moves are concentrated at certain stages of the family cycle, when the household is expanding or contracting, though such moves may have to be delayed if the market, or the household's financial position, prohibit them at certain times. Moving homes is often costly, particularly for home-owners; for flat and apartment renters, on the other hand, it is usually much less traumatic. The latter are generally much more mobile, therefore, and the dwellings to which they move their relatively few belongings also tend to change hands much more frequently. Since flats are often clustered into certain parts of the city, this means that those areas also have the highest mobility rates.

The neighbourhoods of origin and destination for individual moves are usually very similar in their socio-economic, demographic and housing characteristics. Occasional moves, such as the young couple proceeding from a central city flat to a new outer suburban bungalow, qualify this generalization. They also deviate from another general tendency, that for moves to be over relatively short distances. This latter spatial bias results from a combination of several factors. Since most households are moving within a certain neighbourhood type, they have no need to move far, unless they have a clear desire to live in another part of town. Furthermore, their amount of information about various areas is probably negatively related to distance from their home of origin. As many new homes are selected through personal contacts, and these almost certainly have a distance bias, vacancies heard-about tend to decline with increasing distance. Even those heard-of that are some distance away are less likely to be selected, since, although able to assess the suitability of the dwelling, the prospective buyers are usually less able to assess its social environment. Finally, householders have ties to an area, to its residents, and their friends among them, and perhaps to the physical environment also, all of which they are loathe to leave. Hence the search for a 'new' dwelling, often guided by a real estate agent with his own opinion of the 'right' district for a household and perhaps with his own 'market area', tends to be spatially restricted and usually close to the initial home base (Brown and Holmes 1971).

One further spatial bias to intra-urban migrations has been propounded but not yet completely accepted: a sectoral bias (Adams 1969). This suggests that most people's movements are constrained within a certain sector of the city, so that when they move they search within that sector only, for the same reasons that their hunt is constrained by distance – an uneven spatial distribution of information. As well as being tied to a certain sector, it is also suggested that moves are generally outwards within it, largely because most people move to houses of a better quality than those presently occupied, usually new houses, which means moving outwards (in terms of the Burgess and Hoyt models). The relevance of these sectoral and directional biases has been disputed, though much of the debate concerns techniques (Donaldson 1973). Although migrations of suburban households may generally be biased in terms of distance, direction, and sector, however, it is already clear that renters of central-city flats generally display only the first bias in their many moves. This may be a feature of their particular housing market, or it may be more widely applicable.

Permanent migrations are one type of intra-urban spatial behaviour, forming only a very small proportion of the total amount of movement. Some of the other behaviours – shopping, commuting and various social interactions – have already been discussed and distance biases indicated. Other behaviours have been researched and many are summarized by Michelson (1970). The degree to which different groups interact within different spaces has been shown in studies of Belfast (Boal 1969, 1971). Roman Catholic and Protestant families living near to each other differ not only in the newspapers they read and the football teams they support, but also in the shops and bus stops that they patronize, and the areas in which they visit friends and search for spouses. People of different socio-economic status (but no religious differences) occupying adjacent estates also live completely separate social lives, never overlapping at neighbourhood churches, schools or social clubs. The same is often true for ethnic groups living cheek-by-jowl, as in inner Chicago (Suttles 1969), where various Italian, Negro, Mexican and Puerto Rican populations have their own stores and churches (of the same religion). Finally, as a further example of distance biases in the flows of information, Cox (1969) has suggested how political behaviour, expressed in the ballot box, is influenced by the local social environment.

Conclusion: the system as an inter-dependent whole

Apart from the introductory, largely theoretical, section, this chapter has looked at the three major components of the intra-urban system in isolation from each other. Inter-relationships have been stressed where they were necessary for full description of the relevant component system, as in the links between neighbourhood type and shopping centre provision, but the total workings have not been outlined. The main interactions among the components are movements, however, especially movements of people, and these have frequently been mentioned. Over-all models of the city, especially those related to land-use and transport planning, indeed focus on these interactions, using information on land uses to predict traffic flows, and thereby suggest the best future arrangement of the various components (Berry and Horton 1970). Other land uses, such as public utilities and open spaces, also enter these equations as significant generators of trips. Geographical research is now beginning to focus more on these components, relatively minor often in terms of proportion of the total land occupied, but perhaps much more important in the total operation of the system.

The morphology of most intra-urban spatial structures has developed in a laissez-faire environment; the recent advent of stringent planning controls usually channels development more firmly along the path it would have taken, land use zones and new transport routes generally following rather than producing the market demand. The patterns that emerge, then, provide one alternative, which may be relatively efficient in total, given the cumulative inertia and initial advantage of previous investments. Several recent writers, however, have questioned whether these relatively efficient systems are the best for society: both Jacobs (1961) and Sennett (1970), for example, have argued strongly for the spatial mixing, not segregation, of various population groups, though the ends which they seek to achieve vary markedly.

Social justice is a concept at which most societies aim. In the past few years, several writers have associated this with an allied concept of spatial justice (Pahl 1971; Harvey 1971). Thus not only do they ask whether society offers equal opportunity for all social groups, but also whether equal opportunities are offered to groups living in different parts of the system. (The social and spatial groups may, of course, be closely related.) Is the provision of parks and of freeways fairly even over the whole city, or are certain areas favoured? And are the favoured areas those in which the city's rulers live? Do the inner suburban poor have equal access to cheaper shopping facilities as the richer outer suburbanites, or are the latter more favoured by the large supermarkets offering 'specials' and discounts? Unfortunately few answers to these questions are yet available. Much work focuses on the problems of Negroes in United States cities, constrained to inner ghettos and having to travel much longer distances than their white counterparts to obtain outer suburban jobs (Berry and Horton 1970), or being unequally treated in terms of medical provision (Earickson 1970). Just as cities in the developing world have been characterized as having 'dual economies' with few benefits trickling down to the underprivileged, usually underemployed majority, so it might be claimed that similar 'dual economies' are features of 'developed' urban systems, with minorities suffering social injustices because of spatial constraints on their behaviour. Perhaps, then, a major alteration of the morphology of the intra-urban system may eventually be achieved by planning for the average citizen's welfare (and the variation in this) rather than for total efficiency, which tends to distribute its benefits unequally.

4 The international system

Whereas the discussions of systems of places within countries and of intra-urban systems have had a wealth of theoretical and empirical statements on which to draw, this chapter on the international system suffers from a paucity of sources. Geographers have tended to neglect the operation of this, the most extensive of spatial systems, hence the relatively speculative nature of statements made here. Nevertheless, the same basic framework is used, of a system whose component parts can be differentiated on structural, behavioural and demographic criteria and between which distance is an important influence upon patterns of organization and change. Most attention in this chapter is given to the structural dimension of the system, to the pattern of world trade, since it has been more fully researched.

World trade patterns

The growth of urban and regional economies is characterized in chapter 2 as an amalgamation of two processes, import replacement and export diversification, and Thompson's (1965) model of the stages of urban growth is based on these. A similar conceptualization can be applied to the world trading system. Countries specialize in the production of certain goods and services and exchange these, through the international economy, for the products of others. The specialization results from comparative advantages, from the superiority of various places in the performance of certain tasks. It is open to debate, however, as to whether this comparative advantage emerges 'naturally', because of differences in countries' endowments, or whether it is in fact largely an imposed pattern determined by the activities of some states only. Opinion favours the latter interpretation (would New Zealand be a major world producer of sheep meats and dairy produce if it had not been settled from Britain?), though obviously natural resource endowments are relevant to the development of certain comparative advantages.

Whatever the basis for the development of trading patterns, the relevance of the systems approach to their study is clear. In such investigation, however, there is one major difference from the study of the operation of a national space economy. Within modernized countries, the flow of goods is virtually unconstrained by non-economic factors, even in a federal system such as Australia (Smith 1963); note, however, the barrier effect which interprovincial borders have on interaction in Canada (Mackay 1958). As Pounds (1971) and Thoman and Conkling (1967, chapter 7) have described, there is a wide range of institutional intrusions into the free-flow of commodities between countries which must be considered in studies of trade. Nevertheless, there is also evidence that the same principles employed in the rest of this book also apply to world trade, so the institutional arrangements are viewed here as extra independent variables influencing the system's operation. In this sense, the socio-political milieu is perhaps more closely related to the operation of international spatial system than is the case of the more laissez-faire urban systems discussed in earlier chapters.

COMPONENTS OF THE SYSTEM

Thompson's model proposes an urban growth sequence broken into five stages of increasing economic complexity. It has been adapted here for the study of the role of nation states in the international system, with the five stages collapsed to three.

(1) *The export concentration stage*, in which a country's role in the world economy is the provision of only one or a few commodities, usually primary or only semi-processed. Earnings from these are used to purchase a wide range of imports.

(2) *The mature stage*, during which the import replacement process accelerates and a wider range of industries for local consumption is established. As the breadth of industrial (including service industries) expertise increases, so may the range of exports, although considerable dependence on a few staples is likely to remain. The range of imports is unlikely to change markedly; if anything it will increase, since the more complex economy will generate a greater spectrum of demand, for raw materials, or partly processed goods, especially.

(3) *The metropolitan stage*, characterized by the export diversification process as the advancing economy finds itself more able to sell its factories' products on the world market. Again, the import pattern is unlikely to change overall, although the composition may alter considerably. (In order to concentrate on its new export lines, for example, a country may import what it formerly produced for itself.)

Although presented as stages it may not be possible to categorize each country precisely on what is essentially a continuum of development.

Michaely (1962) has studied the export and import concentration of various countries in 1954 according to a 150-commodity trade classification (see also Thoman and Conkling 1967) which suggests positions in the above sequence. His coefficient of commodity concentration uses the formula

$$Cjx = 100\sqrt{\sum\left(\frac{Xij}{Xj}\right)^2}$$

where Xij = country j's exports (imports) in commodity i
$\quad\quad\quad Xj$ = country j's total exports (imports)
$\quad\quad\quad Cjx$ = the desired coefficient

This ranges between 100 and $100/\sqrt{n}$ (in this case $100/\sqrt{150} = 8 \cdot 2$); the closer the index is to 100, the more concentrated the country's trade into certain categories.

Table 4.1 shows the resulting indices for the 44 countries. Export concentration is clearly typical of a great many countries, whereas import concentration is much less. The import concentration index is also much less variable, suggesting that—as the above model proposed—differences in export concentration are more crucial in determining the position of countries in the sequence. The countries in the upper half of the table are therefore in the export concentration stage, with the expected characteristics—except for the Netherlands Antilles whose high level of import concentration results from its function as a crude oil importer and refined product exporter. At the foot of the table are those countries in the metropolitan stage, while in between are states such as Australia which can be categorized as mature, having a somewhat wider range of exports than many, plus a considerable amount of home industry (the latter point, of course, has to be inferred from other data). Division of the countries according to their level of development (whether or not per capita income exceeded $300 US in 1954) and size (whether or not the population exceeded 10 million which might be considered a threshold for large-scale import replacement) produces the following mean indices of export concentration.

	Developed	Underdeveloped	All
Large	21·1	57·9	40·6
Small	39·1	52·0	43·4
All	31·1	55·8	41·9

Clearly, development level is closely associated with a country's role in the system, as is its size among the more developed nations.

A large proportion of countries, mostly of the underdeveloped world, are in the export specialization stage, therefore, contributing only a few products to the world system. (Thoman and Conkling 1967, give more detailed and recent data.) Few countries have moved far beyond this stage, importing the wide range of commodities in

TABLE 4.1   *Concentration in world trade, 1954*

| | Commodity Concentration | | Geographic Concentration | |
	Exports	Imports	Exports	Imports
Mauritius	99·6	23·7	77·6	46·5
Netherlands Antilles	93·7	83·0	33·6	79·7
Colombia	85·0	23·9	79·8	64·2
Egypt	84·2	18·6	26·0	26·0
Ghana	83·5	21·4	47·9	50·5
Iceland	80·3	19·1	29·1	30·8
Burma	74·4	21·8	47·6	39·6
Trinidad and Tobago	72·7	30·5	50·0	47·5
Thailand	68·3	21·4	43·5	35·7
Rhodesia and Nyasaland	63·6	20·2	58·4	55·5
Panama	62·8	16·8	95·5	67·0
Honduras	62·7	18·4	78·0	69·4
Brazil	61·2	23·7	41·5	37·4
Costa Rica	60·5	18·8	62·4	60·4
Australia	50·8	27·0	41·2	49·5
Malaysia and Singapore	49·8	25·4	26·0	38·2
Nigeria	49·3	24·3	74·0	51·3
Kenya–Uganda	49·0	22·9	38·8	52·5
Greece	46·2	19·3	34·1	31·1
Tanganyika	44·6	22·8	39·5	47·2
Indonesia	41·7	24·5	38·3	31·4
Turkey	39·7	23·5	29·3	28·4
Ireland	38·3	16·5	89·7	56·7
Finland	38·1	19·2	34·0	26·8
Mexico	35·0	26·0	73·7	81·2
Libya	34·1	18·9	47·9	45·6
Spain	31·4	23·2	29·5	28·4
Argentina	28·7	23·8	32·1	26·8
Sweden	28·8	18·5	28·4	30·4
Austria	27·7	19·7	31·8	39·7
Denmark	27·1	18·5	41·8	37·0
Belgium–Luxemburg	25·5	15·5	29·4	28·6
Norway	25·5	23·4	28·0	25·9
Canada	24·9	18·0	63·9	74·9
Japan	24·8	26·4	24·0	38·1
Portugal	24·7	19·1	27·4	28·0
Federal Republic of Germany	22·3	15·6	21·0	20·8
Yugoslavia	21·4	26·1	31·3	37·0
Hong Kong	20·6	16·0	27·6	30·0
Italy	20·5	20·7	21·1	24·2
United Kingdom	19·2	16·1	18·7	19·4
United States	18·8	20·5	27·5	27·6
France	18·0	20·4	21·8	20·4
Netherlands	16·9	16·0	27·0	29·6

Source: Michaely (1962, 11–12, 19–20)

which the others specialize among themselves, and trading in return an equally wide range of mainly industrial products. The system, then, appears to be organized around these few metropolitan countries.

Most others specialize in providing a few commodities and importing most of their demands. A few (the mature countries) are intermediate to these two positions.

One further feature highlighted by Michaely (1962) is the influence of accessibility on export concentration. By comparing one group of six countries relatively close to the metropolitan nodes to another of comparable development level but much more remote, he found that members of the first group were much less specialized in their exports. Their mean index of export concentration was 31·7, compared with 51·9 for the latter group. (Countries in the first group were Austria, Greece, Ireland, Italy, Portugal and Spain, whose mean per capita income in the mid 1950s was $300 US; those in the second group – mean income $295 US – were Argentina, Brazil, Colombia, Malaya, Mexico, Panama and Turkey.) This suggests that greater accessibility to the market provides more information and opportunities for industrialists in nearby countries, who can probably employ cheap labour, as well as a distance-biased search by the metropolitan states for commodities they are unable or unprepared to produce themselves. More distant countries do not generally enjoy such advantages, but must specialize to compete on the major markets.

These data all raise the question 'who trades with who?', which can be in part answered by Michaely's coefficients of geographical concentration

$$Gjx = 100\sqrt{\sum\left(\frac{Xsj}{Xj}\right)^2}$$

where Xsj = country j's exports to (imports from) country s
   Xj = country j's total exports (imports)
   Gjx = the required coefficient

Since 44 countries were employed, the theoretical range of this coefficient is 15 – an equal distribution of trade among the 43 others – to 100 – all trade with but one other country.

Table 4.1 shows these coefficients for 1954, again indicating, as for commodity concentration, greater variability and greater average concentration of exports than of imports. The two sets of coefficients (commodity and geographic) are indeed considerably correlated (a rank correlation coefficient of 0·61); countries trading in only a few commodities also tend to trade with few partners. This may appear surprising in that countries specializing in certain production might be expected to distribute it world-wide. In fact, few commodities are the preserve of only one or two countries, especially in the typical primary production of the export specialization stage. Many of these countries are also small, and often tied politically to a metropolitan nation, so their production may not meet one large country's demand.

From these two sets of data in table 4.1 it is possible to suggest a simple model of trade patterns. At the core of this are the metropolitan countries which (a) import certain basic commodities from countries in the export specialization stage; (b) return a wide range of commodities, usually industrial products, to these countries; (c) exchange the products of their more specialized industries among themselves. On the periphery are the export specialization nations, trading a few commodities with a few nations in return for a wider range of imports from the same partners. Finally, there are the mature nations which occupy an intermediate position in this pattern as they move from one extreme of the system towards the other.

THE PATTERN OF INTERACTION

Although indicating that much of the world's trade is concentrated along certain routes, the above data do not indicate which routes. Are there any regularities which suggest the operation of the distance variable as an influence on trading partner choice?

Thoman and Conkling (1967) have presented a number of tables showing the major flows to and from various countries. From these it is possible to verify the metropolitan–periphery concept introduced earlier (see also the data on COMECON trade in Pounds 1971, 306–7). Two other major conclusions are also apparent.

(1) Certain groups of countries tend to trade among themselves only, largely for political motives. The Communist blocs exemplify this; among the Eastern European countries plus the Soviet Union, for example, 66–80% of the trade of each is with other group members. Similar intra-group cohesion is typical of past and present colonial blocs, such as those based on Britain, and more especially on France, Spain, and Portugal.

(2) Within trading groups, nations tend to trade most with their nearer neighbours.

More detailed research on the relationships between trade volume and intervening distance has been reported by Yeates (1969) who applied the well-known gravity formula (see p. 33) to the total trade relations (exports and imports combined) of six nations. A reasonable fit (multiple R varying from 0·46 to 0·73) was obtained with the equation

$$\log T_{ij} = \log a + b \log N_j - \log D_{ij}$$

where
N_j = gross national income of country j
D_{ij} = distance between countries i and j
T_{ij} = trade between countries i and j

Thus, the larger the country (measured by N_j), distance being held constant, the greater the amount of trade. Since the data were transformed to logarithms, both of these relationships have a power form: as the size of partner increases, trade increases at an increasing rate; as its distance away increases, the rate of trade decline decreases. It is of interest that both the regression coefficients are highest for Canada (table 4.2.), presumably because of its considerable trade with the neighbouring USA, and that the distance exponent is smallest for Britain, a country which trades very widely around the world. The low coefficient on the GNI variable for France probably reflects its considerable trade with its many small, ex-colonies.

Further exploration with this model incorporated the other conclusion drawn from Thoman and Conkling's data, the importance of trading groups. Introduction of a dummy variable (c) into the equation

TABLE 4.2 *The gravity model applied to 1964 international trade flows to and from selected countries*

Country	Intercept (a)	Regression Coefficients (b)		Correlation Coefficient (R)
		log National Income	log Distance	
Sweden	5·65	1·28	−1·47	·65
Canada	6·40	1·69	−1·94	·73
France	8·52	0·34	−0·84	·49
South Africa	3·07	1·65	−1·21	·56
Italy	5·84	0·99	−0·68	·57
United Kingdom	5·73	0·72	−0·24	·46

Source: Yeates (1969, 401)

for UK trade, with a Commonwealth, or one-time Commonwealth country being given a value of one whereas all others were coded zero resulted in the equation*

$$\log T_{ij} = 6·2259 + 0·8651 \log N_j - 0·6071 \log D_{ij} + 1·4931C$$

where   C is the dichotomous variable

and an increase in the multiple R from 0·46 to 0·73. This indicates that, comparing two countries of equal size and distance from the UK, an average Commonwealth country conducted about seven times more trade with Britain than did a non-Commonwealth country. McConnell (1971) has computed similar equations for the trade of Israel and Lebanon. These also show that the countries trade with large and nearby partners, especially with those in certain political blocs, within which the size and distance variables still operate.

The most detailed study of interaction patterns in the world trading system is Linneman's (1966). His basic equation involved the following independent variables

the Gross National Product of both origin and destination
the population of both origin and destination

the distance between origin and destination
preferences among the UK and its associate countries, France and
its associate countries, and Portugal and Belgium with their associate countries.

Again, the data were logarithmically transformed.* The equation was
fitted to several data sets, which varied according to the inclusion or
exclusion of very small and zero trade volumes, for both exports and
imports as well as total trade. In most cases a multiple correlation
coefficient of approximately 0·8 resulted, with all of the independent

TABLE 4.3   *Regression coefficients for the distance variable: international*
trade flows by country, 1959

Regression coefficient (b)	Number of cases	
	Exports	Imports
Positive, less than 0·2	2	
Negative, less than 0·2	2	3
0·2–0·39	2	2
0·4–0·59	6	7
0·6–0·79	9	10
0·8–0·99	19	14
1·0–1·19	11	10
1·2–1·39	5	13
1·4–1·59	6	4
1·6–1·79	4	4
1·8–1·99		1
2·0–2·19	1	
2·2–2·39		3
2·4–2·59	2	

Source: Linneman (1966, 90)

variables contributing significantly to the final equation. A further
variable was then included which measured the complementarity*
between the exports from the origin and the imports at the destination.
Not surprisingly, this too proved significant as an explanation of trade
volumes, though the increase in R was generally slight.

Linneman also computed separate equations for each country (he
used 1959 data). In the present context, major interest is in the regression coefficients for the distance variable, whose frequency distribution is given in table 4.3. Only in two cases – both for exports –
was this value positive, and most cases ranged from −0·6 to −1·19.
Further tabulations suggested that the highest values, indicating the
most rapid distance-decay patterns, were for countries at 'the geo-

graphical periphery of the world economy' (Linneman 1966, 91), with the exceptions of New Zealand and Uruguay. This was particularly so for export trade, with peripheral countries sending more of their products to their nearest neighbours. More generally, however, Linneman's tabulations suggested that distance exponents were lowest for the more developed countries.

Recently, Pederson and Stohr (1971) have suggested inter-commodity variations in the influence of distance on trade within South America. Raw materials are widely traded, presumably because they are moved by relatively cheap coastal shipping and their purchase is largely by big companies possessing detailed market information. The much smaller flows of manufactured goods, mainly moved by inland transport modes, generally occur between adjacent countries; Pederson and Stohr relate this to the small size of the firms involved, the poverty of information flows concerning markets, and the more expensive transport.

These twin influences of trade blocs and distance are summarized in Russett's (1967) classification of trading nations, which proved very stable between 1954 and 1963. Factor analysis suggested nine groups of 'choosers' and 'chosen':

(1) Countries of Central America and the Caribbean trading with the USA and industrial Western Europe.

(2) South American nations plus those European countries culturally linked to them.

(3) British Commonwealth countries, especially those of the Caribbean.

(4) The British sphere of influence, including many Commonwealth countries, plus the major European industrial states. Britain itself was the link between these two.

(5) The French sphere plus metropolitan France's neighbours.

(6) The EEC and its neighbours.

(7) The COMECON.

(8) The Arab League.

(9) Non-Communist Asia.

Clearly these groups overlap, as some of the metropolitan countries are in several of them (France, for example, is in both 5 and 6). Some countries were not classifiable, and each group contained members which would not generally be anticipated. The existence of strong patterns of territoriality is suggested by most groups as a major influence which over-rides the distance variable: no country of the British Caribbean is in the first group; none from Latin America is in the second. Political and cultural systems thus interlock with distance to influence trade flows, therefore, although, despite the evidence quoted above, Russett (1967, 143) claims that the influence of distance may be exaggerated.

EMERGENCE OF THE SPATIAL SYSTEM

The present world trade system largely comprises a North Atlantic-centred metropolitan core and a world-wide periphery, within which are a few countries, such as Australia, Japan and South Africa, that are at the mature stage of the process outlined earlier and are moving, especially rapidly in the Japanese case, towards the establishment of new metropolitan nodes. Development of this system has largely been outwards from Britain, a country favoured by various aspects of its social and economic structure to take the lead in economic development (Parsons 1971). There are two related strands in this spread process.

(1) Expansion of the metropolitan core.

(2) Articulation of a wider system by the acquisition of new areas to the export specialization periphery organized around the core.

The first of these processes has itself followed a two-pronged expansion. One has been a distance-biased, spread effect (similar to that observed in local diffusion trends); the other has used the maritime connections between Britain and certain other countries. Within Europe, political and institutional changes have been necessary before countries could follow the nineteenth-century British model (Parsons 1971; Kuznets 1966); for example, 'the new economic epoch became evident only after steam railroads with their far-reaching effects on size of markets and industrial technology had proven themselves' (Kuznets 1966, 440). Both shared antecedents and strong communications within the core were the precursors of development. Research has suggested that this spread was strongly associated with distance from London. Godlund's (1960) map of the dates of railway opening in various parts of Europe supports this argument, as does a recent study by Casetti, King and Williams (1972). They mapped income per capita in European capitals in 1860, 1880, 1900 and 1913 and related these variations to time and distance from London. Multiple regression analysis* suggested the validity of both variables (producing a high multiple correlation coefficient of 0·85), with a velocity of spread for each income isopleth of 21 km per year.

Beyond Europe, the core has been joined by certain ex-British dependencies, plus Japan, while in Europe it has expanded into Russia. The former countries were those that were permanently settled, in contrast to those which provided only temporary homes for a European elite in which colonial expropriation has been the more typical pattern of economic expansion. The institutional conditions for the drive to metropolitan status, via the import replacement and export diversification processes and assisted by increasing size, emerged at different dates, hence the present differing levels of development among these countries. Of the present metropolitan and mature stage countries, only Japan and China are exceptions to this generalization (though Israel is a special case); for these two, Kuznets (1966, 475) suggests that economic expansion followed late nineteenth century aggressive contacts with the West.

The second process by which the world system grew, enlargement of the periphery, has also contained a major element of distance-bias (especially if distance is measured in transport costs, thereby emphasizing the relatively short-hauls across oceans). Peet (1969) has suggested that a von Thünen-like zonation of agricultural land uses, expanding over time, has developed around a 'world urban-industrial

TABLE 4.4 *Average distance transported for British agricultural imports, 1830–1913 (excluding Ireland)*

Commodity	Average Distance to London				
	1831–5	1856–60	1871–5	1891–5	1909–13
Fruit and Vegetables	nil	325	535	1150	1880
Live Animals	nil	630	870	3530	4500
Butter, Cheese, Eggs	262	530	1340	1610	3120
Feed Grains	860	2030	2340	3240	4830
Flax and Seeds	1520	3250	2770	4080	3900
Meat and Tallow	2000	2900	3740	5050	6250
Wheat and Flour	2430	2170	4200	5150	5950
Wool and Hides	2330	8830	10,000	11,010	10,900

Source: Peet (1969, 295)

nucleus', formerly based on London, but now 'occupying much of Western Europe and north-eastern North America' (Peet 1969, 290). Early in the nineteenth century, therefore, London's food supply area was the rest of Britain, plus Ireland and the Baltic. As the century proceeded, however, the average distance various commodities were moved to London increased (table 4.4). The expansion was not concentric because of the preference given to British-oriented territories and the cheapness of water-borne transport. In general, however, the outer edge – the wheat frontier – proceeded across the USA and Canada, and thence to the Southern Hemisphere colonies. A similar, more local expansion has produced a zonal pattern of supply of horticultural and dairy produce for Britain (Chisholm 1962).

The dominance of water transportation modes in this expansion of the periphery has meant that many places had for some time (and still have in much of Africa and South America) a dual economy. One part of this, on the seaboard or near to other navigable water, is integrated with the world trading system; the other represents an isolated, more self sufficient economy (see Johnson 1970, for an example of this in nineteenth-century Ireland). Thus Blainey (1966) described the initial settlement of Australia as a series of 'limpet' ports; in the 1820s around Sydney, overland transport costs were such that wheat could not be profitably moved more than 58 km. Only the railways allowed expansion of agriculture into the interior. Even with time-space convergence as far advanced as it now is, Brookfield and Hart (1971) have characterized present-day Melanesia as 'von Thünen's outermost ring'; its main products are industrial crops, low yielding in weight after processing, which until recently could only be profitably grown 'within 10 km of the sea'.

Expansion of the industrial economy has followed similar routes into the outlying British-oriented territories. In Canada, for example, over half of all manufacturing plants are owned by US companies, who locate their factories in response to two main factors, accessibility to the market and distance from the US headquarters (Ray 1965, 1971). This pattern is strongest in Quebec and the Atlantic Provinces, and has led Ray (1971, 399) to conclude that 'the geography of regional development in Canada . . . [has] been shaped by a process of industrial interactance with United States metropolitan centers'.

CONTEMPORARY REGIONAL DEVELOPMENTS

In their drive away from export specialization and towards industrial maturity, many countries have recently rejected the type of development process outlined in the previous paragraphs. Instead they prefer local, publicly-owned, import-replacement industrialization. For many countries, however, progress in this way is inhibited by their small size and inability to achieve substantial scale economies for fledgling industries. Attempts are being made to widen markets, therefore, by economically, if not politically, integrating groups of countries, following the model set in the metropolitan countries with Benelux and then the EEC.

In terms of the model of world trade outlined in this chapter, such international economic integration should reduce the commodity and geographic concentration, especially in exports, for the countries involved. Balassa (1966) and Conkling and McConnell (1971) have shown this to have been so in the EEC and Nordek communities. Economic integration has also been attempted by five countries which have formed the Central American Common Market (CACM), and the effects of this on the trade specialization of Costa Rica, El Salvador, Guatemala, Honduras, and Nicaragua have recently been investigated (McConnell and Conkling 1972). Since CACM became effective in 1961, data for 1953, 1960 and 1967 have been analysed (Conkling and McConnell, 1972). Between the last two dates, commodity specialization in exports – initially very high – fell considerably, especially for Guatemala, El Salvador and Costa Rica, which countries possessed the largest cities and the best infrastructure for industrial developments. Guatemala provides the most extreme case; its commodity export concentration coefficient (see above, p. 93), computed over ten categories only, fell from 96·8 in 1953 to 61·6 in 1967. Imports were diversified too, but less so, since the five economies are not really complementary so little of their increased range of primary products is in strong demand within the community.

Within national space-economies, it has been suggested that time-space convergence generates processes of centralization, deconcentration and specialization. It is widely accepted that similar processes, especially the first, are typical of the world system, although most countries aim to prevent the growing hegemony of a few nations. Hence the development of trading blocs may reduce centralization, replacing it by more deconcentration, characterized by specialization in inter-bloc trade but the converse within blocs.

The behavioural and demographic dimensions

Within national and regional areas, the urban system articulates not only the exchange of goods and services, but also the flows of peoples and ideas. Similar processes apparently operate in the wider international system. The countries of the metropolitan core act as the innovation hearthlands, for example, and people and ideas flow from these along the established channels of communication – biased both by distance and by cultural links – to the mature countries and those on the international periphery.

THE DIFFUSION OF ATTITUDES AND IDEAS

Although it is easy to postulate such a system of world interaction, there is little available evidence with which to test it. Nevertheless, general observation clearly indicates that most social changes originate from but a few countries, more especially from their major cities. London, New York and Paris are internationally accepted as world leaders in the introduction of a wide range of behaviours, from fashions to music, hair styles to drug use, student protest to social attitudes. And even where an innovation may have had a prior existence elsewhere, as with the 'rediscovery' of many Indian religions, including yoga, during the 1960s, this is usually adopted first in one of the metropolitan cities mentioned before being diffused to the

H

rest of the world, along the established routes (even to those places where institutional controls attempt to halt such a spread). The reasons for this concentration of social innovation in the world's metropolitan centres are the same as those quoted in chapter 2: size and the availability of scale economies; a relatively anonymous population; a willingness and ability to innovate; and the clustering of initiative.

The spread of such innovations is exemplified by Wagner's (1971) description of the areal expansion of the Latin language, albeit from another, earlier metropolitan core. As the Romans assimilated the various alien tribes, a five-stage process of language development occurred.

(1) Initial spread of the relatively uniform usage.
(2) Immediate succeeding onset of local diversification.
(3) Explicit efforts at standardization.
(4) Complete divorce of the artificially preserved language from the descendant vernaculars.
(5) (a) The death of the standard as a living language.
    (b) New standards developed.

In its totality, this sequence may not apply to all spread processes – stage 3 may be irrelevant in many cases, for example. But the spread of the English language, of many religions, notably Christianity, and of British legal and constitutional systems are just a few examples of phenomena which probably fit Wagner's model.

The agents of spread for such innovations were people and books, and the dissemination of the innovations was a function of the extent and speed of the transport system, Today, the agents of spread are much more the mass media, through which communication of ideas to all parts of the world is very rapid. Since the speed of introduction is now so fast, and adoption so ephemeral, it is doubtful whether local vernaculars will develop in many spheres. Australian Rules and American Football, for example, would probably not have developed as hybrids under present conditions of communication. Instead, no sooner has a new fashion from London or Paris reached the Antipodes than another is on its way.

The spread of Latin, of Roman Catholicism, and of many other innovations in past centuries represented an almost wholesale transference of cultural norms. Recent decades have seen a similar spread of what might be termed the 'development syndrome', of the idea of material progress, although this has been like the spread of religions to Africa, with a variety of development ideologies from separate metropolitan hearths competing to establish their spheres of influence. Such ideologies have spread in the same way as the other examples given in this chapter, along the culturally and distance-biased routes. Like the

European agricultural hinterland, they have their flexible outward moving frontiers, which are occasionally breached, as with Cuba in 1962, when one ideology attempts to establish an exclave in another sphere of influence, from which to initiate further spread processes.

DEMOGRAPHIC CHANGE

It is the demographic changes associated with the spread of the 'development syndrome' which are most frequently documented. Dixon (1971), for example, has charted a trend, spreading among 'Western' countries since 1945, of earlier and more universal marriage. Concurrently with this, 'Eastern' countries are introducing constraints

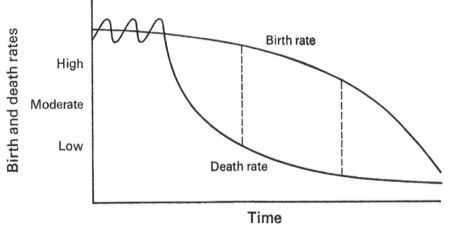

fig. 4.1 *The demographic transition model. The three divisions of the model conform to those depicted in fig. 4.2.*

on early marriage. Eventually, she suggests, the two initially very different sets of attitudes may converge.

One of the main demographic features of economic change, neither cause nor effect, but probably a necessary precondition (Goode 1963), is the changing role of the family in society. One aspect of this is changing fertility norms, which are a key element in, for example, the arguments for zero population growth. The spread of these changing norms through a national system was outlined in chapter 2; a study of a similar spread pattern (Chung 1971) over the entire world in the twentieth century is reported here.

This study was based on the position of each country at various dates on the well-known Demographic Transition (fig. 4.1), which is a general representation of the relation of birth and death rates during a period of economic development. This has three main stages. In the first, both birth and death rates are high, so that population growth is slight. During the second, both rates fall markedly, but since the decline in death rate usually predates and outpaces that in the birth rate the stage is characterized by a period of very rapid population growth. Finally, both rates stabilize at a relatively low level, so that growth rates are retarded. In addition, Chung recognized an intermediate type within stage two, in which both birth and death rates were moderate and population growth not too great (see key to fig. 4.2).

Each country was categorized according to this classification for each five-year period from 1905–9 to 1955–9. Three of the main maps are reproduced here (fig. 4.2). In 1911–13, only a few countries of Western Europe (including Britain), plus North America and Australasia were at stage 3; a few others, mostly European but also South Africa and Uruguay, were in stage 2, and the rest were in stage 1. Low growth rates were general, therefore. (Note that this refers to natural growth only; no data are included on migration.) During the 1920s and 1930s most European countries passed through to stage 3 (France and Spain were notable exceptions), as did Argentina and Uruguay. Apart from these two, and Japan, however 'the rest of Latin America, Africa, and Asia seem to be asleep under the demographic blanket of high fertility and high mortality' (Chung 1971, 234). During the 1940s, the transition spread east through Europe, and, by 1945–9 (fig. 4.2B), several Latin American countries had also entered stage 2. Finally, in the period 1955–9 (fig. 4.2C) all of Europe and the Soviet world had reached stage 3, while much of Africa, Asia and Latin America were either entering, or well into, the explosive growth characteristics of stage 2.

The main agent of the process outlined in the maps of fig. 4.2, especially the entry of many countries into stage 2, has been the widespread adoption of death control technology, unparalleled by any significant spread and adoption of the complementary birth control technology. This differs from the countries already in stage 3, where birth control has expanded rapidly in recent years while the already low mortality rates have displayed little overall variation since 1950–3. On the other hand, the causes of death have altered quite considerably, and the metropolitan countries may now be initiating the spread of new 'killers', often associated with changing life styles, which could have important consequences on countries' courses through the demographic transition. Tuberculosis, influenza and pneumonia may no longer be fatal diseases, but in the countries of and around the metropolitan core, cancer, diabetes, cirrhosis of the liver, suicide, and motor accident death rates are all rapidly increasing.

MIGRATION

Zelinsky (1971) has postulated, though he has presented little supporting evidence for, a further suite of processes which should parallel the

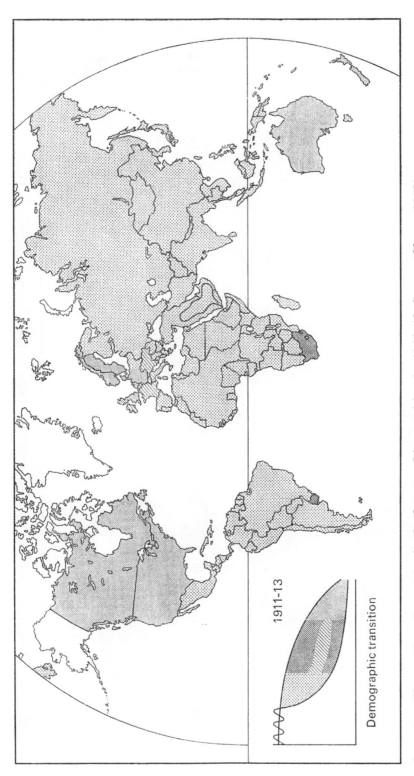

1911–13

Demographic transition

fig. 4.2a *The position of countries in the demographic transition in 1911–13. Source: Chung (1971).*

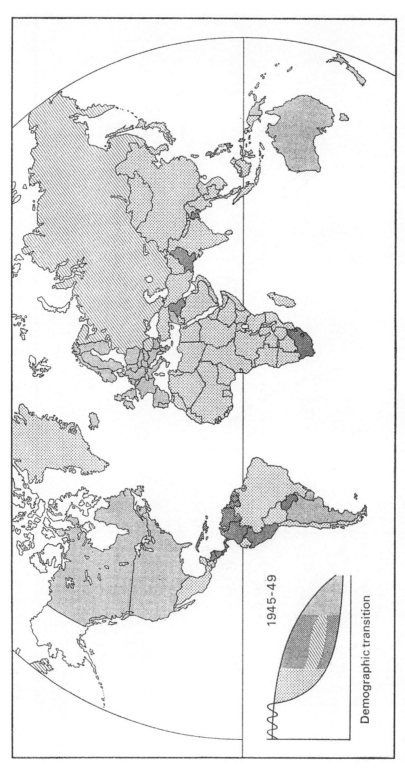

fig. 4.2b *The position of countries in the demographic transition in 1945–9. Source: Chung (1971).*

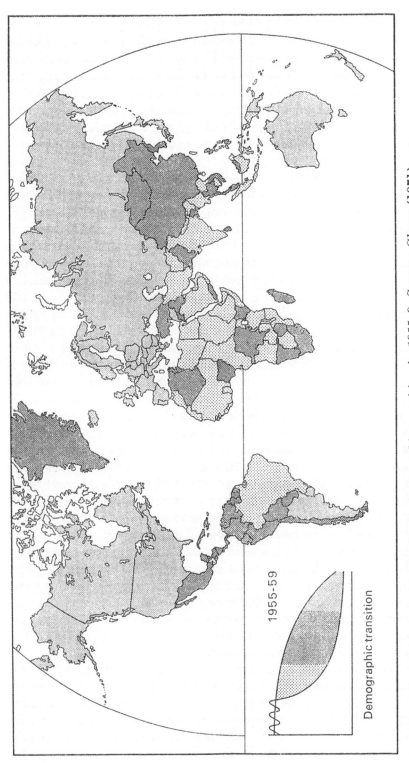

1955-59

Demographic transition

fig. 4.2c   The position of countries in the demographic transition in 1955–9. Source: Chung (1971).

demographic transition; he terms this the mobility transition (fig. 4.3). Its main components refer to the volumes of international and internal migration which, as his speculative diagrams suggest, should vary as a society progresses towards its final stage of development. At this time, as a result of technological advances and associated time-space convergence, much of the former volume of migration will be replaced by circulation and communications. In some ways, this model is an elaboration of that presented by Gibbs (1963) for the process of population concentration.

If empirical tests confirm Zelinsky's speculations, they will indicate that attitudes to migration, and propensities to migrate, will change as countries proceed through the development process, just as their rates of natural increase are presumed to do. One problem with both the demographic and the mobility transition models is that their relationship to the spread of modernization has not been fully specified. Indeed, it would seem that in recent decades the onset of the transitions has preceded, not paralleled, other aspects of modernization, notably industrialization. As a result, the death rate has fallen much more rapidly in, for example, Latin America than it did in Europe a century earlier, without compensating changes in fertility levels. This has produced a very different occupational structure in the former countries, with a much greater emphasis on tertiary employment (see Mehta 1961, on Burma). With the early introduction of capital-intensive rather than labour-intensive industrialization many of the developing countries have become 'overurbanized' (implying that too many of their people live in the cities). Whether this is an apt term, or whether twentieth-century urbanization is a different process from that of the preceding century, is a topic of some debate (Sovani 1964). In any case, the changing form of the relationships with modernization and with the various aspects of demographic change requires detailed specification in the testing of Zelinsky's hypotheses.

International migration is one aspect of world demography which should illustrate the basic principles of the organization of spatial systems, although, like world trade, it is more constrained by institutional arrangements than is the flow of people within any one state. Nevertheless, it is possible to identify three major migration patterns similar to those in the typology outlined in chapter 2 (p. 48).

(1) At the lowest level, there is labour migration, much of it temporary, even seasonal. Most of it is over relatively short distances, like the nineteenth-century Irish harvest migrations to Britain, and oriented towards the nearest opportunities. In some cases, specialized labourers may be moved long distances on certain contracts (Italian tunnellers to power schemes in New Zealand, for example), but movements like those of the Mexican *braceros* to the adjacent United

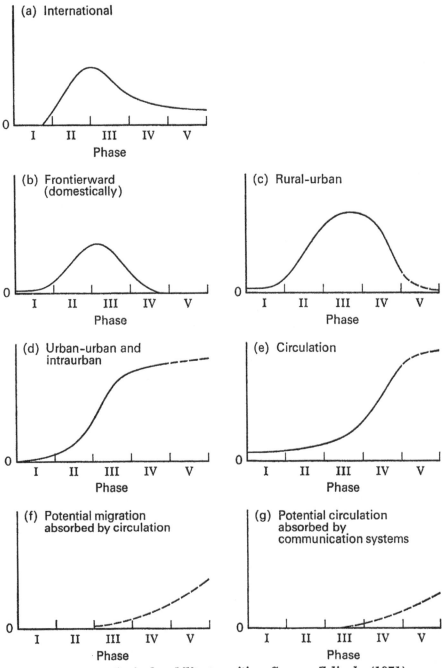

fig. 4.3 *The hypothesized mobility transition. Source: Zelinsky (1971).*

States are much more typical. International tourist flows may display some of these characteristics, too (Zelinsky and Williams 1970).

Labour migration is of considerable volume in some parts of the world, notably Europe. During 1965, for example, nearly three million workers were registered as having moved from six southern European countries (Magee 1971). France, Switzerland and West Germany were the main recipients of these flows (table 4.5); each tended to draw most

TABLE 4.5   *Labour migration in Europe, 1965 (in thousands)*

Destination	Greece	Italy	Portugal	ORIGIN Spain	Turkey	Yugoslavia	Total
Austria	2	3	—	2	5	19	31
Belgium–Luxemburg	8	95	2	32	17	—	155
France	5	340	103	346	8	20	822
Netherlands	2	8	1	16	6	1	34
West Germany	187	372	14	183	133	65	954
Sweden	—	5	—	—	—	5	13
Switzerland	7	500	1	79	4	4	595
United Kingdom	—	—	—	—	—	—	100
Total	225	1500	125	675	180	125	2830

Source: Magee (1971, 179)

of its workers from the nearest source, while outmigrants in general moved to the nearest opportunity. Thus Switzerland drew almost entirely from Italy, and, given the absence of further jobs there, most Greeks, Turks, and Yugoslavs migrated to West Germany. France's main migration field was the Iberian Peninsula.

(2) The intermediate level comprises permanent migration for long-term gain as well as the short-term objectives more characteristic of the first type. In line with the model outlined in chapter 2, such moves should be 'up' the 'hierarchy' of countries, in terms of their position on the metropolitan-export specialization continuum. The massive migrations to the United States over the last century or so typify this, as, to a lesser extent, do those to Australia of the present century, the more recent exodus of West Indians and Asians to Britain, and of Pacific Islanders to New Zealand. More detailed research may suggest that such flows are to the nearest possible alternative, given the general cultural constraints plus the matrix of institutional barriers and encouragements to migration existing at the time.

(3) Finally, there is the equivalent of what Taeuber and Taeuber (1964) called the 'inter-metropolitan circulation of elites'. In the international context this contains such elements as the moves made by academics among the metropolitan countries, the international transfer of company executives to branch companies in developing countries,

and the transfer of civil servants among international organizations. The latter is likely to increase markedly in volume with such developments as the enlargement of the EEC.

One topic not developed in this chapter concerns the spatial system of international relations. This has been studied by several political scientists in recent years, notably Russett (1967), who has investigated the major characteristics of such patterns as diplomatic relations, international conflict, and voting in the United Nations. Spatially contiguous groupings of countries are frequently identified in such analyses, suggestive of a distance-bias in information flows, but there has been little modelling of the system's operations along the lines suggested here. Such work may, however, isolate clear spatial structures of the type outlined in this chapter for other aspects of the international system, thereby adding to the range of materials for which the spatial structures theme is relevant.

Conclusions

As stressed in the introductory comments, this chapter is much more speculative than the two preceding it, largely because geographers have displayed little interest in the operation of the world as a spatial system. No justifiable conclusions can be presented here; rather, all that can be hoped for is that the present set of arguments will stimulate a little more research into the general area. Enough is known, however, for the validity of viewing the world in the same spatial terms as the study of urban systems to have been demonstrated. As with all analogies, the commonalities can be over-stressed, for the change of scale involves moving from a relatively homogeneous to a largely heterogeneous social, political and cultural system. But the commonalities would seem to be there, as a useful basis for the geographical study of international systems.

References

ABU-LUGHOD, J. L. (1969) Testing the theory of social area analysis: the ecology of Cairo, Egypt. *Am. Sociol. Rev.* 34, 198–212.

ADAMS, J. S. (1969) Directional bias in intra-urban migration. *Econ. Geogr.* 45, 302–23.

AHMAD, Q. (1965) *Indian Cities: Characteristics and Correlates.* Univ. of Chicago, Dept. Geogr. Res. Pap. 102. Chicago.

ALEXANDER, J. W. *et al.* (1958) Freight rates: selected aspects of uniform and nodal regions. *Econ. Geogr.* 34, 1–18.

ALONSO, W. (1964) *Location and Land Use.* Cambridge, Mass.

ALONSO, W. (1969) Urban and regional imbalances in economic development. *Econ. Dev. & Cult. Change* 17, 584–95.

BALASSA, B. (1966) Tariff reductions and trade in manufactures among the industrial countries. *Am. Econ. Rev.* 56, 466–72.

BARNUM, H. G. (1966) *Market Centers and Hinterlands in Baden-Wurttemberg.* Univ. of Chicago, Dept. Geogr. Res. Pap. 103. Chicago.

BEAVER, S. H. (1961) Geography and technology. *Adv. of Sci.* 18, 315–27.

BEGG, H. M. (1972) Remoteness and the location of the firm. *Scot. Geogr. Mag.* 88, 48–52.

BERRY, B. J. L. (1961) City size distribution and economic development. *Econ. Dev. & Cult. Change* 9, 673–87.

BERRY, B. J. L. (1963) *Commercial Structure and Commercial Blight.* Univ. of Chicago, Dept. Geogr. Res. Pap. 85. Chicago.

BERRY, B. J. L. (1964a) Approaches to regional analysis: a synthesis. *Ann. Assoc. Am. Geogr.* 54, 2–11.

BERRY, B. J. L. (1964b) The case of the mistreated model. *Prof. Geogr.* 16 (3).

BERRY, B. J. L. (1966) *Essays on Commodity Flows and the Spatial Structure of the Indian Economy.* Univ. of Chicago, Dept. Geogr. Res. Pap. 111. Chicago.

BERRY, B. J. L. (1967a) Cities as systems within systems of cities. *Pap. Reg. Sci. Assoc.* 13, 147–63.

BERRY, B. J. L. (1967b) *The Geography of Market Centers and Retail Distribution.* Englewood Cliffs, N.J.

BERRY, B. J. L. (1970a) City size and economic development, in JAKOBSON, L. and PRAKESH, V. (eds.) *Urbanization and National Development.* California. 111–56.

BERRY, B. J. L. (1971a) (ed.) Comparative factorial ecology. *Econ. Geogr.* 47.

BERRY, B. J. L. (1971b) The geography of the United States in the year 2000. *Trans. Inst. Brit. Geogr.* 51, 21–54.

BERRY, B. J. L. (1972) (ed.) *City Classification Handbook.* New York.

BERRY, B. J. L. and HORTON, F. E. (1970) *Geographic Perspectives on Urban Systems.* Englewood Cliffs, N.J.

BERRY, B. J. L. and REES, P. H. (1969) The factorial ecology of Calcutta. *Am. J. Sociol.* 74, 445–91.

BERRY, B. J. L. and SPODEK, H. (1971) Comparative ecologies of large Indian cities. *Econ. Geogr.* 47, 266–85.

BERRY, B. J. L. *et al.* (1966) *Central Place Theory: A Review and Bibliography.* Reg. Sci. Res. Inst. Philadelphia.

BLAIKIE, P. M. (1971) Spatial organization of agriculture in some North Indian villages. *Trans. Inst. Brit. Geogr.* 52, 1–40 and 53, 15–30.

BLAINEY, G. (1966) *The Tyranny of Distance.* Melbourne.

BLECHYNDEN, K. (1964) An economic base analysis of Hamilton. *N. Z. Geogr.* 20, 122–37.

BOAL, F. W. (1969) Territoriality on the Shankhill–Falls divide, Belfast. *Ir. Geogr.* 6, 30–50.

BOAL, F. W. (1971) Territoriality and class; a study of two residential areas in Belfast. *Ir. Geogr.* 6, 229–48.

BOAL, F. W. and JOHNSON, D. B. (1965) The functions of retail and service establishments on commercial ribbons. *Can. Geogr.* 9, 154–69.

BRITTON, J. N. H. (1967) *Regional Analysis and Economic Geography: a case study of manufacturing in the Bristol region.* London.

BROOKFIELD, H. C. with HART, H. D. (1971) *Melanesia: a Geographical Interpretation of an Island World.* London.

BROWN, L. A. (1968) *Diffusion Processes and Location.* Reg. Sci. Res. Inst. Philadelphia.

BROWN, L. A. GOLLEDGE, R. G. and ODLAND, J. (1970) Migration, functional distance and the urban hierarchy. *Econ. Geogr.* 46, 472–85.

BROWN, L. A. and HOLMES, J. H. (1971) Search behaviour in an intra-urban migration context: a spatial perspective. *Env. & Plan.* 3, 307–26.

BRUSH, J. E. and BRACEY, H. E. (1955) Rural service centers in Southwestern Wisconsin and Southern England. *Geogr. Rev.* 45, 559–69.

BURGESS, E. W. (1924) The growth of the city; an introduction to a research project. *Proc. Am. Sociol. Soc.* 18, 85–97.

BURGHARDT, A. F. (1971) A hypothesis about gateway cities. *Ann. Assoc. Am. Geogr.* 61, 269–85.

BURNLEY, I. H. (1972) Ethnic settlement formation in Wellington-Hutt. *N. Z. Geogr.* 28, 151–70.

CARLSSON, G. (1966) The decline of fertility: innovation or adjustment process. *Pop. Stud.* 20, 149–74.

CASETTI, E., KING, L. J. and ODLAND, J. (1971) The formalization and testing of concepts of growth poles in a spatial context. *Env. & Plan.* 3, 337–82.

CASETTI, E., KING, L. J. and JEFFREY, D. (1971) Structural imbalance in the U.S. urban economic system, 1960–1965. *Geogr. Anal.* 3, 239–55.

CASETTI, E., KING, L. J. and WILLIAMS, F. (1972) Concerning the spatial spread of economic development. *Internat. Geogr.* 2, 897–99.

CHINITZ, B. (1965) Contrasts in agglomeration: New York and Pittsburgh. *Am. Econ. Rev.* 61, 279–89.

CHISHOLM, M. (1962) *Rural Settlement and Land Use.* London.

CHISHOLM, M. (1971) Freight transport costs, industrial location, and regional development, in CHISHOLM, M. and MANNERS, G. (eds.) *Spatial Policy Problems of the British Economy.* Cambridge, 213–44.

CHRISTALLER, W. (1933) *Die Centralen Orte in Suddeutschland.* Jena. (Available translated by BASKIN, C. W. (1966) Englewood Cliffs, N.J.)

CHUNG, R. (1971) Space–time diffusion of the transition model, in DEMKO, G. J., ROSE, H. M. and SCHNELL, G. A. (eds.) *Population Geography: A Reader.* New York.

CLARK, W. A. V. (1968) Consumer travel patterns and the concept of range. *Ann. Assoc. Am. Geogr.* 58, 386–96.

CLARK, W. A. V. and RUSHTON, G. (1970) Models of intra-urban consumer behavior and their implications for central place theory. *Econ. Geogr.* 46, 486–97.

COHEN, Y. (1972) *The Diffusion of Planned Shopping Centers in the United States.* Univ. of Chicago, Dept. Geogr. Res. Pap. 135. Chicago.

COLEMAN, J. S., KATZ, E., and MENDEL, H. (1966) *Medical Innovation: A Diffusion Study.* Indianapolis.

CONKLING, E. C. and MCCONNELL, J. E. (1971) Integration and commodity trade specialization; the case of Nordek, in HAMILTON, F. E. I. (ed.) *Festschrift: Arthur E. Moodie.* Northwestern Univ. Stud. Geogr. 18. Evanston, 1–42.

COX, K. R. (1969) The voting decision in a spatial context. *Progress in Geography* 1, 81–117.

COX, K. R. and DEMKO, G. J. (1968) Agrarian structure and peasant discontent in the Russian Revolution of 1905. *E. Lakes Geogr.* 3, 3–20.

DAHMS, F. A. (1971) Urban residential structure: some neglected factors. *N.Z. Geogr.* 27, 130–50.

DALY, M. T. (1968) Residential location decisions: Newcastle, New South Wales. *Aust. & N.Z. J. Sociol.* 4, 18–35.

DARROCH, A. G. and MARSTON, W. G. (1971) The social class basis of ethnic residential segregation: the Canadian case. *Am. J. Sociol.* 77, 491–510.

DAVIE, M. R. (1938) The pattern of urban growth, in MURDOCH, G. P. (ed.) *Studies in the Science of Society.* New Haven.

DAVIES, R. L. (1968) Effects of consumer income differences on the business provisions of small shopping centres. *Urb. Stud.* 5, 144–64.

DAVIES, R. L. (1969) Effects of consumer income differences on shopping movement behaviour. *Tijdschr. Econ. Soc. Geogr.* 60, 111–21.

DAVIS, K. (1965) The urbanisation of the human population. *Scient. Am.* 213 (3) 40–53.

DICKINSON, R. E. (1934) Markets and market areas of East Anglia. *Econ. Geogr.* 10, 172–82.

DIXON, R. (1971) Explaining cross-cultural variations in age at marriage and proportion never marrying. *Pop. Stud.* 25, 215–34.

DONALDSON, B. (1973) An empirical investigation into the concept of sectoral bias in the mental maps, search spaces and migration patterns of intra-urban migrants. *Geografiska Ann.* 55B.

DUNCAN, B. and LIEBERSON, S. (1970) *Metropolis and Region in Transition.* California.

DUNCAN, O. D. and DUNCAN, B. (1955) Residential differentiation and occupational stratification. *Am. J. Sociol.* 50, 493–503.

DUNCAN, O. D. and LIEBERSON, S. (1959) Ethnic segregation and assimilation. *Am. J. Sociol.* 64, 364–74.

DUNCAN, O. D. et al. (1960) *Metropolis and Region.* Baltimore.

EARICKSON, R. (1970) *The Spatial Behavior of Hospital Patients.* Univ. of Chicago, Dept. Geogr. Res. Pap. 123. Chicago.

ELIAS, N. and SCOTSON, J. L. (1965) *The Established and the Outsiders.* London.

FELLER, I. (1973) Determinants of the composition of urban inventions. *Econ. Geogr.* 49, 47–58.

FLORIN, J. W. (1971) The diffusion of the decision to integrate. *S.E. Geogr.*

GARNER, B. J. (1966) *The Internal Structure of Retail Nucleations.* Northwestern Univ. Dept. Geogr. Res. Pap. 12. Evanston.

GIBBS, J. P. (1963) The evolution of population concentration. *Econ. Geogr.* 39, 119–29.

GILSON, M. (1970) The changing New Zealand family: a demographic analysis, in HOUSTON, H. S. (ed.) *Marriage and the Family in New Zealand*. Wellington, 41–65.

GODDARD, J. B. (1967) Changing office location patterns within Central London *Urb. Stud.* 4, 276–85.

GODDARD, J. B. (1968) Multivariate analysis of office location patterns in the city centre: a London example. *Reg. Stud.* 2, 69–85.

GODDARD, J. B. (1971) Office communications and office location: a review of current research. *Reg. Stud.* 5, 263–80.

GOLLEDGE, R. G. (1963) The New Zealand brewing industry. *N.Z. Geogr.* 19, 7–24.

GOLLEDGE, R. G. (1970) *Process Approaches to the Analysis of Human Spatial Behavior*. Ohio State Univ. Dept. Geogr. Discussion Pap. 16. Columbus, Ohio.

GOLLEDGE, R. G., RUSHTON, G. and CLARK, W. A. V. (1966) Some spatial characteristics of Iowa's dispersed population and their implications for the grouping of central place functions. *Econ. Geogr.* 42, 261–72.

GOODE, W. J. (1963) Industrialization and family change, in HOSE-LITZ, B. F. and MOORE, W. E. (eds.) *Industrialization and Society*. London, 237–55.

GOODWIN, W. (1965) The management center in the United States. *Geogr. Rev.* 55, 1–16.

GOULD, P. R. (1970a) Tanzania 1920–63. The spatial impress of the modernization process. *Wld. Pol.* 22, 149–70.

GOULD, P. R. (1970b) *Spatial Diffusion*. Assoc. Am. Geogr. Commis. on Coll. Geogr. Res. Pap. 4. Washington, D.C.

GOULD, P. R. and SPARKS, J. (1969) The geographical context of human diets in Southern Guatemala. *Geogr. Rev.* 59, 58–82.

GOULD, P. R. and WHITE, R. R. (1969) The mental maps of British school leavers. *Reg. Stud.* 2, 161–82.

GRIFFIN, D. W. and PRESTON, R. E. (1966) A restatement of the transition zone concept. *Ann. Assoc. Am. Geogr.* 56, 339–50.

HÄGERSTRAND, T. (1966) *Innovation Diffusion as a Spatial Process* (trans. A. Pred). Chicago.

HAGGETT, P. (1965) *Locational Analysis in Human Geography*. London.

HAGGETT, P. (1972) *Geography: A Modern Synthesis*. New York.

HALL, P. G. (1962) *The Industries of London since 1861*. London.

HALL, P. G. (1966) (ed.) *Thünen, J. H. von: Isolated State*. Oxford.

HARRIES, K. D. (1971) Ethnic variations in Los Angeles business patterns. *Ann. Assoc. Am. Geogr.* 61, 736–43.

HARRIS, C. D. (1954) The market as a factor in the localization of industry in the United States. *Ann. Assoc. Am. Geogr.* 44, 315–48.

HARVEY, D. W. (1971) Social processes, spatial form and the re-distribution of real income in an urban system, in CHISHOLM, M., FREY, A. E. and HAGGETT, P. (eds.) *Regional Forecasting.* London, 270–300.

HAY, A. M. and SMITH, R. H. T. (1970) *Interregional Trade and Money Flows in Nigeria 1964.* Ibadan.

HAYNES, K. E. *et al.* (1973) Intermetropolitan migration in high and low opportunity areas: indirect tests of the distance and inter-vening opportunities hypotheses. *Econ. Geogr.* 49, 68–73.

HODDER, B. W. (1961) Rural periodic day markets in part of Yoruba-land. *Trans. Inst. Brit. Geogr.* 29, 149–60.

HODDER, B. W. (1965) Some comments on the origins of traditional markets in Africa South of the Sahara. *Trans. Inst. Brit. Geogr.* 36, 97–105.

HODGE, G. (1966) The prediction of trade center viability in the Great Plains. *Pap. Proc. Reg. Sci. Assoc.* 15, 87–115.

HOYT, H. (1939) *The Structure and Movement of Residential Neighbor-hoods in American Cities.* Federal Housing Administration. Washing-ton, D.C.

HUNKER, H. L. and WRIGHT, A. J. (1963) *Factors of Industrial Location in Ohio.* Ohio State Univ. Columbus.

HUNTER, J. M. and YOUNG, J. A. C. (1971) Diffusion of influenza in England and Wales. *Ann. Assoc. Am. Geogr.* 61, 637–53.

ISARD, W. (1969) *General Theory.* Cambridge, Mass.

ISARD, W. and KUENNE, R. E. (1953) The impact of steel upon the Greater New York–Philadelphia industrial region: a study of agglomeration projection. *Rev. Econ. & Statistics,* 35, 289–301.

JACOBS, J. (1961) *The Death and Life of Great American Cities.* New York.

JACOBS, J. (1971) *The Economy of Cities.* New York.

JANELLE, D. G. (1968) Central place development in a time–space framework. *Prof. Geogr.* 20, 5–10.

JANELLE, D. G. (1969) Spatial reorganization: a model and concept. *Ann. Assoc. Am. Geogr.* 59, 348–64.

JEFFERSON, M. (1939) The law of the primate city. *Geogr. Rev.* 29, 226–32.

JEFFREY, D. CASETTI, E. and KING, L. J. (1969) Economic fluctua-tions in a multiregional setting: a bi-factor analytic approach. *J. Reg. Sci.,* 9, 397–404.

JOHNSON, J. H. (1970) The two 'Irelands' at the beginning of the nineteenth century, in STEPHENS, N. and GLASSCOCK, R. E. (eds.) *Irish Geographical Studies in Honour of E. Estyn Evans.* Belfast, 224–43.

JOHNSTON, R. J. (1966a) Central places and the settlement pattern. *Ann. Assoc. Am. Geogr.* 56, 541–9.

I

JOHNSTON, R. J. (1966b) The distribution of an intra-metropolitan central place hierarchy. *Aust. Geogr. Stud.* 4, 19–33.

JOHNSTON, R. J. (1966c) The location of high status residential areas. *Geografiska Ann.* 48B. 23–35.

JOHNSTON, R. J. (1968a) Railways, urban growth and central place patterns. *Tijdschr. Econ. Soc. Geogr.* 59, 33–41.

JOHNSTON, R. J. (1968b) Population change in Australian small towns. *Rur. Sociol.* 34, 212–18.

JOHNSTON, R. J. (1970) Latent migration potential and the gravity model. *Geogr. Anal.* 2, 387–97.

JOHNSTON, R. J. (1971a) *Urban Residential Patterns: An Introductory Review.* London.

JOHNSTON, R. J. (1971b) Regional development in New Zealand. *Reg. Stud.* 5, 321–31.

JOHNSTON, R. J. (1972a) *Content foci in human geography.* Monash Publ. Geogr. No. 4. Monash Univ. Victoria, Australia.

JOHNSTON, R. J. (1972b) Continually changing human geography: a review of some recent literature. *N.Z. Geogr.* 28, 78–96.

JOHNSTON, R. J. (1972c) Activity spaces and residential preferences: some tests of the hypothesis of sectoral mental maps. *Econ. Geogr.* 48, 199–211.

JOHNSTON, R. J. (1972d) Towards a general model of intra-urban residential patterns: some cross-cultural observations. *Progress in Geogr.* 4, 83–124.

JOHNSTON, R. J. (1973) Social area change in Melbourne 1961–1966: a sample exploration. *Aust. Geogr. Stud.* 11.

JOHNSTON, R. J. and KISSLING, C. C. (1971) Establishment use patterns within central places. *Aust. Geogr. Stud.* 9, 116–32.

JOHNSTON, R. J. and RIMMER, P. J. (1967) A note on consumer behavior in an urban hierarchy. *J. of Reg. Sci.* 7, 161–6.

JOHNSTON, R. J. and RIMMER, P. J. (1969) *Retailing in Melbourne.* Aust. National Univ., Dept. Hum. Geogr.

KARASKA, G. J. (1966) Interindustry relations in the Philadelphia economy. *E. Lakes Geogr.* 2, 80–96.

KEEBLE, D. E. (1968) Industrial decentralization and the metropolis: the North-west London case. *Trans. Inst. Brit. Geogr.* 44, 1–54.

KEEBLE, D. E. (1969) Local industrial linkage and manufacturing growth in outer London. *Tn Plan. Rev.* 40, 163–88.

KEEBLE, D. E. and HAUSER, D. P. (1971, 1972) Spatial analysis of manufacturing growth in outer South-east England. *Reg. Stud.* 5, 229–62, and 6, 11–36.

KEOWN, P. A. (1971) The career cycle and the stepwise migration process. *N.Z. Geogr.* 27, 175–84.

KNOS, D. S. (1968) The spatial distribution of land values in Topeka, Kansas, in BERRY, B. J. L. and MARBLE, D. F. (eds.) *Spatial Analysis: A Reader in Statistical Geography*. Englewood Cliffs, N.J. 269–89.

KRUEGEL, D. L. (1971) Metropolitan dominance and the diffusion of human fertility patterns, Kentucky: 1939–1965. *Rur. Sociol.* 36, 141–56.

KUHN, A. (1966) *The Study of Society: A Multidisciplinary Approach*. London.

KUZNETS, S. S. (1966) *Modern Economic Growth*. New Haven.

LACHENE, R. (1966) Networks and the location of economic activities. *Pap. Reg. Sci. Assoc.* 14, 183–96.

LEE, E. S. (1966) A theory of migration. *Demog.* 3, 47–57.

LINNEMAN, H. (1966) *An Econometric Study of International Trade Flows*. Amsterdam.

LINSKY, A. S. (1965) Some generalizations concerning primate cities. *Ann. Assoc. Am. Geogr.* 55, 506–13.

LLOYD, P. E. (1965) Industrial changes in the Merseyside development area. *Tn Plan. Rev.* 35, 285–98.

LOGAN, M. I. (1964) Suburban manufacturing: a Bankstown case study. *Aust. Geogr.* 9, 223–34.

LOGAN, M. I. (1966) Locational behavior of manufacturing firms in urban areas. *Ann. Assoc. Am. Geogr.* 56, 451–66.

LOSCH, A. (1954) *The Economics of Location*. New Haven.

MCCOLL, R. W. (1969) The insurgent state: territorial basis of revolution. *Ann. Assoc. Am. Geogr.* 59, 613–31.

MCCONNELL, J. E. (1971) A flow algorithm for the trade of small developing nations. *Pennsylvania Geogr.* 9, 10–15.

MCELRATH, D. C. (1968) Societal scale and social differentiation: Accra, Ghana, in GREER, S. MCELRATH, D. C., MINAR, D. W. and ORLEANS, P. W. (eds.) *The New Urbanization*. New York, 33–52.

MACKAY, J. R. (1958) The interactance hypothesis and boundaries in Canada: a preliminary study. *Can. Geogr.* 11, 1–8.

MAGEE, B. A. (1971) Problems of economic development and migration in Southern Europe with special reference to Spain. *Proc. Sixth N.Z. Geogr. Conf.* 178–83.

MANNERS, G. (1964) *The Geography of Energy*. London.

MEHTA, S. K. (1961) A comparative analysis of the industrial structure of the urban labor force of Burma and the United States. *Econ. Dev. & Cult. Change* 9, 164–79.

MEYER, D. R. (1971) *Spatial Variation of Black Urban Households*. Univ. of Chicago, Dept. Geogr. Res. Pap. 129. Chicago.

MICHAELY, M. (1962) *Concentration in World Trade*. Amsterdam.

MICHELSON, W. (1970) *Man in his Urban Environment: A Sociological Approach*. Reading, Mass.

MILLS, E. S. (1972) Welfare aspects of national policy toward city sizes. *Urb. Stud.* 9, 117–24.

MORRILL, R. L. (1970) *Spatial Organization of Society*. Belmont, California.

MORRIS, R. N. (1968) *Urban Sociology*. London.

MOSER, C. A. and SCOTT, W. (1962) *British Towns*. Edinburgh.

MURDIE, R. A. (1965) Cultural differences in consumer travel. *Econ. Geogr.* 41, 211–13.

MURDIE, R. A. (1969) *Factorial Ecology of Metropolitan Toronto 1951–1961*. Univ. of Chicago, Dept. Geogr. Res. Pap. 116. Chicago.

MURPHY, R. E. (1966) *The American City: An Urban Geography*. New York.

MURPHY, R. E., VANCE, J. E. and EPSTEIN, B. J. (1955) Internal structure of the CBD. *Econ. Geogr.* 31, 21–46.

NADER, G. A. (1971) Some aspects of the recent growth and distribution of apartments in the Prairie metropolitan areas. *Can. Geogr.* 15, 307–17.

NEUTZE, G. M. (1965) *Economic Policy and the Size of Cities* Dept. of Econ. Aust. National Univ. Canberra.

NORTHAM, R. M. (1969) Population size, relative location and declining urban centres: conterminous United States 1950–1960. *Ld Econ.* 45, 313–22.

PAHL, R. E. (1971) Poverty and the urban system, in CHISHOLM, M. and MANNERS, G. (eds.) *Spatial Policy Problems of the British Economy*. Cambridge, 126–45.

PARKER, H. R. (1962) Shopping in Liverpool. *Tn Plan. Rev.* 33, 197–223.

PARSONS, T. (1971) *The System of Modern Societies*. Englewood Cliffs, N.J.

PEDERSON, P. O. (1970) Innovation diffusion within and between national urban systems. *Geogr. Anal.* 2, 203–54.

PEDERSON, P. O. and STOHR, W. (1971) Economic integration and the spatial development of South America, in MILLER, J. and GAKENHEIMER, R. A. (eds.) *Latin American Urban Problems and the Social Sciences*. Beverley Hills, 73–103.

PEET, J. R. (1969) The spatial expansion of commercial agriculture in the nineteenth century: a von Thünen interpretation. *Econ. Geogr.* 45, 283–301.

PIRENNE, H. (1925) *Medieval Cities*. Princeton.

POUNDS, N. J. G. (1971) *Political Geography*. New York.

PRED, A. R. (1964a) A typology of manufacturing flows. *Geogr. Rev.* 54, 65–84.

PRED, A. R. (1964b) The intrametropolitan location of industry. *Ann. Assoc. Am. Geogr.* 54, 165–80.

PRED, A. R. (1965a) Industrialization, initial advantage and American metropolitan growth. *Geogr. Rev.* 55, 158–85.

PRED, A. R. (1965b) The location of high value added manufacturing. *Econ. Geogr.* 41, 108–32.

PRED, A. R. (1967) *Behavior and Location, Part I.* Lund.

PRED, A. R. (1968) *Behavior and Location, Part II.* Lund.

PRED, A. R. (1971a) Urban systems development and the long-distance flow of information through pre-electronic U.S. newspapers. *Econ. Geogr.* 47, 498–524.

PRED, A. R. (1971b) Large-city interdependence and the pre-electronic diffusion of innovations in the U.S. *Geogr. Anal.* 3, 165–81.

PRED, A. R. and KIBEL, B. M. (1970) An application of gaming simulation to a general model of economic locational processes. *Econ. Geogr.* 46, 136–56.

QUINN, J. A. (1940) The Burgess zonal hypothesis and its critics. *Am. Sociol. Rev.* 5, 210–18.

RAMSØY, N. R. (1966) Assortative mating and the structure of cities. *Am. Sociol. Rev.* 31, 773–86.

RAVENSTEIN, E. G. (1885) The laws of migration. *J. Roy. Stat. Soc.* 48, 167–235.

RAY, D. M. (1965) *Market Potential and Economic Shadow.* Univ. of Chicago, Dept. Geogr. Res. Pap. 101. Chicago.

RAY, D. M. (1969) The spatial structure of economic and cultural differences: a factorial ecology of Canada. *Pap. Reg. Sci. Assoc.* 23, 7–23.

RAY, D. M. (1971) The location of United States manufacturing subsidiaries in Canada. *Econ. Geogr.* 47, 389–400.

REES, P. H. (1971) Factorial ecology: an extended definition, survey and critique of the field. *Econ. Geogr.* 47, 220–33.

REES, P. H. (1972) Problems of classifying subareas within cities, in BERRY, B. J. L. (ed) *City Classification Handbook.* New York, 265–330.

RICHARDSON, H. W. (1972) Optimality in city size, systems of cities and urban policy: a sceptic's view. *Urb. Stud.* 9, 29–48.

RIMMER, P. J. (1967a) The changing status of New Zealand seaports, 1833–1960. *Ann. Assoc. Am. Geogr.* 57, 88–100.

RIMMER, P. J. (1967b) The search for spatial regularities in the development of Australian seaports. *Geografiska Ann.* 49B, 42–54.

RIMMER, P. J. (1969) *Manufacturing in Melbourne.* Aust. National Univ. Dept. of Hum. Geogr. Canberra.

ROSE, A. J. (1966) Dissent from down under: metropolitan primacy as the normal state. *Pac. Viewp.* 7, 127.

ROSE, W. D. (1969) Manufacturing development policy in New Zealand, 1958–1968. *Pac. Viewp.* 10, 57–77.

RUSHTON, G. (1969) Temporal changes in space-preference structures. *Proc. Assoc. Am. Geogr.* 1, 129–32.

RUSSETT, B. M. (1967) *International Regions and the International System.* Chicago.

SCHNORE, L. F. (1965) On the spatial structure of cities in the two Americas, in HAUSER, P. M. and SCHNORE, L. F. (eds.) *The Study of Urbanization.* New York, 347–98.

SCHWIND, P. M. (1971) *Migrational and Regional Development in the United States,* 1950–1960. Univ. of Chicago, Dept. Geogr. Res. Pap. 133. Chicago.

SCHWIRIAN, K. P. (1972) Factorial ecologies and modernization, in SWEET, D. C. (ed.) *Models of Urban Structure.* Lexington.

SCOTT, A. J. (1971) *Combinatorial Programming, Spatial Analysis and Planning.* London.

SCOTT, P. (1959) The Australian CBD. *Econ. Geogr.* 35, 290–314.

SENNETT, R. (1970) *The Uses of Disorder.* New York.

SHEVKY, E. and BELL, W. (1955) *Social Area Analysis.* Stanford.

SIMMONS, J. W. (1964) *The Changing Pattern of Retail Location.* Univ. of Chicago, Dept. Geogr. Res. Pap. 92. Chicago.

SIMMONS, J. W. (1968) Changing residence in the city: a review of intra-urban mobility. *Geogr. Rev.* 58, 622–51.

SKINNER, G. W. (1964–5) Marketing and social structure in rural China. *J. of Asian Stud.* 24.

SMITH, R. H. T. (1963) Transport competition in Australian border areas. *Econ. Geogr.* 39, 1–13.

SMITH, R. H. T. (1972) (ed.) Spatial structure and process in tropical West Africa. *Econ. Geogr.* 48, 229–355.

SOJA, E. W. (1968) *The Geography of Modernization in Kenya.* Syracuse Univ. Dept. Geogr. Syracuse.

SOVANI, N. V. (1964) The analysis of 'over-urbanization'. *Econ. Dev. & Cult. Change.* 12, 113–22.

SPELT, J. A. (1955) *The Urban Development of South-Central Ontario.* Assen, Netherlands.

STEGMAN, M. (1969) Accessibility models and residential location. *J. Am. Inst. of Planners* 35, 22–9.

STONE, G. P. (1954) City shoppers and urban identification: observations on the social psychology of city life. *Am. J. Sociol.* 60, 36–45.

STOUFFER, S. A. (1940) Intervening opportunities: a theory relating mobility and distance. *Am. Sociol. Rev.* 5, 845–67.

STOUFFER, S. A. (1960) Intervening opportunities and competing migrants. *J. of Reg. Sci.* 2, 1–26.

SUTTLES, G. D. (1969) *The Social Order of the Slum*. Chicago.

SWEETSER, F. L. (1965) Factor structure as ecological structure in Helsinki and Boston. *Acta Sociologica* 8, 205–25.

TAEUBER, K. E. (1968) The effect of income redistribution on racial residential segregation *Urb. Affairs Q.* 3, 5–14.

TAEUBER, K. E. and TAEUBER, A. F. (1964) White migration and socio-economic differences between cities and suburbs. *Am. Sociol. Rev.* 29, 718–29.

TARVER, J. D. (1969) Gradients of urban influence on the educational, employment and fertility patterns of women. *Rur. Sociol.* 34, 356–67.

TARVER, J. D. (1971) Differentials and trends in actual and expected distance of movement of interstate migrants. *Rur. Sociol.* 36, 563–71.

TARVER, J. D. and MCLEOD, R. D. (1970) Trends in the distance of movement of interstate migrants *Rur. Sociol.* 35, 523–33.

TARVER, J. D., TURNER, R. E. and GURLEY, W. R. (1966) *Urban Influences on Oklahoma Farm Population Characteristics and Farm Land Use*. Oklahoma State Univ. Monog. in the Humanities, Social and Biological Sciences, Social Science Series 14. Stillewater, Oklahoma.

TEGSJO, B. and OBERG, S. (1966) Concept of potential applied to price formation. *Geografiska Ann.* 48B, 51–8.

THOMAN, R. S. and CONKLING, E. C. (1967) *The Geography of International Trade*. Englewood Cliffs, N.J.

THOMPSON W. R. (1965) *A Preface to Urban Economics*. Baltimore.

TIMMS, D. W. G. (1971) *The Urban Mosaic: Towards a Theory of Residential Differentiation*. Cambridge.

VANCE, J. E. (1970) *The Merchant's World*. Englewood Cliffs, N.J.

VAPNARSKY, C. A. (1969) On rank size distributions of cities: an ecological approach. *Econ. Dev. & Cult. Change* 17, 584–95.

WAGNER, P. L. (1971) *Environments and Peoples*. Englewood Cliffs, N.J.

WARREN, K. (1966) Steel pricing, regional economic growth and public policy. *Urb. Stud.* 3, 185–99.

WEBBER, M. J. (1972) *The Impact of Uncertainty on Location*. Canberra.

WHEATLEY, P. (1967) Proleptic observations on the origins of urbanism, in STEEL, R. W. and LAWTON, R. (eds.) *Liverpool Essays in Geography*. London, 315–45.

WILSON, M. G. A. (1968) Changing patterns of pit location on the New South Wales coalfields. *Ann. Assoc. Am. Geogr.* 58, 78–90.

WIRTH, L. (1938) Urbanism as a way of life. *Am. J. Sociol.* 44, 1–24.

WISE, M. J. (1949) On the evolution of the jewellery and gun quarters in Birmingham. *Trans. Inst. Brit. Geogr.* 15, 57–72.

WOOD, P. A. (1969) Industrial location and linkage. *Area* 2(2), 32–9.

YEATES, M. H. (1965) Some factors affecting the spatial distribution of Chicago land values, 1910–1960. *Econ. Geogr.* 42, 57–70.

YEATES, M. H. (1969) A note concerning the development of a geographic model of international trade. *Geogr. Anal.* 1, 399–404.

ZELINSKY, W. (1971) The hypothesis of the mobility transition. *Geogr. Rev.* 61, 219–349.

ZELINSKY, W. and WILLIAMS, A. V. (1970) On some patterns in international tourist flows. *Econ. Geogr.* 46, 549–67.

ZIPF, G. K. (1949) *Human Behavior and the Principle of Least Effort.* Reading, Mass.

Glossary of terms used

Complementarity   The degree to which the exports of one place meet the import requirements of another. Complementarity is thus a prerequisite for trade.

Demand Elasticity   The amount of variation in demand for a product relative to price variations.

Distance-Decay   A broad term describing any pattern or process which becomes less intense with increasing distance from a set point.

Dummy Variable   Dummy variables are used to introduce categorical measurements (usually of the form yes:no) to regression equations, which generally use only continuous measurements. In most cases, observations recording 'yes' are coded 1; 'no' are coded zero. The regression coefficient for a dummy variable indicates the higher (if it is +) or lower (if it is −) average value for a 'yes' observation as compared with a 'no' observation, holding constant all other variables in the equation.

Establishment   In central place theory, an establishment is a separate unit, usually a building (shop or office), within which a single business may offer a range of functions. A tobacconist-cum-barber shop is therefore a single establishment.

Function   In central place theory, a function is a separate type of business. There can be more than one function conducted in a single establishment. Similarly, a town may perform a variety of functions.

Industrial Mix   A term describing the composition of a place's industries, usually measured by its employment structure.

Information Field   The area from which an individual, resident at a given point, usually obtains information. It is the area he knows best, and within which most of his activities are located. Most information fields display distance-decay characteristics: the individual has more information about a place near to his home than about one further away.

1*

Logarithmic Transformation For regression analyses (see definition of multiple regression) it is necessary for relationships to be linear: that is, the rate of increase of values on the two variables must be constant. Where this is not so, logarithmic transformation may alter a curvilinear to a linear regression, because of its basis in ratios rather than intervals between numbers. The basic equation

$$Y = a + bX$$

may be transformed to:

$$(1)\ Y = a + b \log X$$

when the rate of increase in Y remains constant as the rate of increase in X increases;

$$(2)\ \log Y = \log a + bX$$

when the rate of increase in Y increases while the rate of increase of X remains constant; and

$$(3)\ \log Y = \log a + b \log X$$

when the rates of increase in both variables increase together. Thus when the variable is untransformed, the increase rate is arithmetic – adding a constant value; when it is transformed, the rate is geometric – multiplying by a constant value. Equation (1) therefore, may read $Y = 0 + 5 \log X$: interpreted this says that each time the logarithm of X increases by 1·0 (or X is multiplied by 10), 5 times that value is added to the value of Y.

Multiple Regression This technique investigates continuous linear relationships between pairs of variables. A simple regression

$$Y = a + bX$$

states that the value of Y is equal to the value of X, multiplied by b, plus a constant value a. The a and b values must be determined from a data set giving values of X and Y for a set of observations. The equation is then used to predict values of Y from known values of X. The success of this fitting is measured by r, the correlation coefficient: r values vary from $+1·0$ to $-1·0$. The closer r is to $\pm 1·0$ the better the fit: a negative value indicates that as X increases, Y decreases; a positive value indicates that X and Y increase together. The b value indicates the slope of the regression. The larger the value of b, the more rapid the rate of increase in Y relative to the increase in X: b values may also be positive or negative. The a value shows where the regression line (with slope b) intercepts the Y axis: it is the average value of Y when X equals zero.

Multiple regression extends this system to any number of variables, with equations such as

$$\dot{Y} = a + b_1 X_1 + b_2 X_2 + b_3 X_3$$

The goodness of fit is measured by the multiple correlation coefficient R, which is always positive, varying between 0·0 (no fit) and 1·0 (perfect fit). The value of R^2 is interpreted as the amount of variation in Y – the dependent variable – which can be statistically accounted for by the variation in the Xs – the independent variables. The b values indicate the slope of the relationship between the relevant indepedent variable and Y, when the effects of all the other independent variables in the equation are held constant. The a value is interpreted as before.

Tests of significance are often used to inquire whether any one variable makes an unambiguous contribution to the goodness-of-fit, i.e. whether the value of R, and thus the predictive power of the equation, is improved by the inclusion of that variable. Such significance tests are based on the principles of statistics, much of which depend on the nature of the sample data used.

Normative Model Such a model describes the 'best' solution to a locational pattern, given a set of assumptions. Central place theory is thus a normative model of the size, spacing and functions of settlements, within the set of assumptions on which the theory is based. Normative models may be considered as optima, against which reality may be compared, to assess the efficiency of the latter in terms of spatial organization.

Principal Components Analysis (and Factor Analysis) It is possible to measure the characteristics of places in a wide variety of ways. Many of these variables may be redundant, however, since they either (a) measure the same thing, or (b) are perfectly correlated in their distributions over the set of observations. Factor analysis is a family of techniques (of which principal components analysis is one of the most popular among geographers) that seeks such redundancy in sets of variables. The techniques mostly work on matrices of correlation coefficients – measures of goodness-of-fit among pairs of variables – and reduce these to sets of hybrid variables which represent combinations of the initial set. In this way, the number of variables is reduced, hopefully to a more manageable (and also interpretable) set of descriptions. It is also possible to obtain scores of these descriptions, relating each individual observation to the new variable as against the combination of original variables which it represents.

Reciprocal Transformation The reciprocal of a number, n, is obtained by dividing it into 1·0 (1/n = reciprocal n). It is useful in studying relationships where, for example, the distance-decay is very rapid initially, but then declines considerably. (For example, land values fall very quickly over the first few hundred yards from a city centre, but the rate of decline slows thereafter, until it is barely perceptible over short distances.) A reciprocal transformation for the distance variable is often able to transform such a relationship to a linear one, thereby allowing the use of regression techniques. (Note that a reciprocal transformation alters the sign of the b value in a distance-decay curve from negative to positive.)

Rural–Urban Continuum   The concept of a rural-urban continuum suggests a regular transition from the two polar types of rural and urban, for example in fertility rates. Lowest fertility may be a feature of the largest cities in a system; highest fertility of the most isolated rural areas. Between these two, the larger the place, the lower its fertility.

Appendix

The metric system; conversion factors and symbols

In common with several other text-book series *The Field of Geography* uses the metric units of measurement recommended for scientific journals by the Royal Society Conference of Editors.* For geography texts the most commonly used of these units are:

Physical quantity	Name of unit	Symbol for unit	Definition of non-basic units
length	metre	m	basic
area	square metre	m^2	basic
	hectare	ha	$10^4 m^2$
mass	kilogramme	kg	basic
	tonne	t	$10^3 kg$
volume	cubic metre	m^3	basic-derived
	litre	l	$10^{-3} m^3$, 1 dm³
time	second	s	basic
force	newton	N	kg m s^{-2}
pressure	bar	bar	$10^5 Nm^{-2}$
energy	joule	J	$kgm^2 s^{-2}$
power	watt	W	$kgm^2 s^{-3} = Js^{-1}$
thermodynamic temperature	degree Kelvin	°K	
customary temperature, t	degree Celsius	C	$t/°C = T/°K - 273·15$

Fractions and multiples

Fraction	Prefix	Symbol	Multiple	Prefix	Symbol
10^{-1}	deci	d	10	deka	da
10^{-2}	centi	c	10^2	hecto	h
10^{-3}	milli	m	10^3	kilo	k
10^{-6}	micro	μ	10^6	mega	M

* Royal Society Conference of Editors, *Metrication in Scientific Journals*, London, 1968.

The gramme (g) is used in association with numerical prefixes to avoid such absurdities as mkg for μg or kkg for Mg.

Conversion of common British units to metric units

Length

1 mile = 1·609 km	1 fathom = 1·829 m
1 furlong = 0·201 km	1 yard = 0·914 m
1 chain = 20·117 m	1 foot = 0·305 m
	1 inch = 25·4 mm

Area

1 sq mile = 2·590 km²	1 sq foot = 0·093 m²
1 acre = 0·405 ha	1 sq inch = 645·16 mm²

Mass

1 ton = 1·016 t	1 lb = 0·454 kg
1 cwt = 50·802 kg	1 oz = 28·350 g
1 stone = 6·350 kg	

Mass per unit length and per unit area

1 ton/mile = 0·631 t/km	1 ton/sq. mile = 392·298 kg/km²
1 lb/ft = 1·488 kg/m	1 cwt/acre = 125·535 kg/ha

Volume and capacity

1 cubic foot = 0·028 m³	1 gallon = 4·546 l
1 cubic inch = 1638·71 mm³	1 US gallon = 3·785 l
1 US barrel = 0·159 m³	1 quart = 1·137 l
1 bushel = 0·036 m³	1 pint = 0·568 l
	1 gill = 0·142 l

Velocity

1 m.p.h. = 1·609 km/h
1 ft/s = 0·305 m/s
1 UK knot = 1·853 km/h

Mass carried X distance

1 ton mile = 1·635 t km

Force

1 ton-force = 9·964 kN
1 lb-force = 4·448 N
1 poundal = 0·138 N
1 dyn = 10⁻⁵ N

Pressure

1 ton-force/ft² = 107·252 kN/m²
1 lb-force/in² = 68·948 mbar
1 pdl/ft² = 1·488 N/m²

Energy and power

1 therm	= 105·506 MJ	1 Btu	= 1·055 kJ
1 hp	= 745·700 W(J/s)	1 ft lb-force	= 1·356 J
	= 0·746 kW	1 ft pdl	= 0·042 J
1 hp/hour	= 2·685 MJ	1 cal	= 4·187 J
1 kWh	= 3·6 MJ	1 erg	= 10^{-7} J

Metric units have been used in the text wherever possible. British or other standard equivalents have been added in brackets in a few cases where metric units are still only used infrequently by English-speaking readers.

Index